TELCO GLOBAL CONNECT Vol : 2

Author : Sadiq Malik

Copyright © 2015 All rights reserved.

ISBN-13: 978-1515008002

DEDICATION

" For Dr Jean Grey...my Phoenix Hurting "

This book is dedicated to the profitable growth of the $1+ trillion Telco industry as the need for transformation and search for new business models becomes imperative. Through a series of insightful articles the book explores various technology, commercial and financial themes that underscore the need to digitise infrastructure and adopt innovative technical, commercial and financial strategies across the Telco value chain.

" Sometimes when you innovate you make mistakes. It is best to admit them quickly and get on with improving your other innovations "

Steve Jobs

PREFACE

While Telcos will always be under pressure to maintain revenue and profitability targets for core/legacy services, they will also need to find new ways of growing revenue. Much of the growth will come from noncore/new businesses such as mobile services, Big Data and cloud. However, they also will need to take several steps such as setting up a new Line of Business, incentivizing innovation and collaboration more effectively throughout the organization, reskilling their workforces from selling communications products and services to selling business-sensitive solutions, and developing more sophisticated pricing models in order to enable them to bundle legacy and new services.

By engaging in collaborative optimization, leveraging network intelligence and evolving dumb pipes into smart ones, Telcos can pragmatically lay a foundation for a profitable and sustainable converged network—one that can support internal and externally procured services that customers will continue to demand over time.Telcos need to stop thinking about networks as discrete islands because In a converged digital world that's no longer how they are being used.

Telcos are being disrupted because the basis of competition in mobile has fundamentally changed from "reliability and scale of networks" to "choice and flexibility of services", representing transition from "mobile telephony" to "mobile computing". Telcos must move their innovation focus from technologies (be it HTML5, NFC, IMS, VoLTE, M2M or RCS-e) to ecosystems. That requires a much better understanding of how ecosystems are engineered, and how ecosystems absorb and amplify innovation.

TABLE OF CONTENTS

Each chapter addresses a particular theme and the articles therein can be ready in any sequence.

Chapter 1 : The Rise of the MVNO

This Chapter highlights knowledge kernels that expose the evolving role of the MVNO business model. In developing countries MVNO's are just starting as incumbent Telcos realise that cannot reach every customer segment and perhaps they should spend more effort on upgrading their networks. In developed markets MVNO need to develop innovative commercial models to monetise from their selected customer niches. MVNO's must consider how the platform fits into their specific business model, measuring cost of ownership with potential for harnessing time-to-market advantages to grow subscriptions and generate revenue. Regulators have a key role to play in the growth of MVNOs. After a first wave of market liberalisation, in which alternative operators began competing with the historical incumbent using their own spectrum and infrastructure, MVNOs can enter in a second wave – ensuring greater competition and consumer choice in their markets.

1.1	Kickstart MVNO's : A look into the WiFi First and Advertising business model	11
1.2	MVNO 's in Africa : time to rethink and retool the business model	17
1.3	MVNO Industry Africa : Learning from the Winners	23

1.4	MVNO Success : Its all about doing Maths	27
1.5	Enterprise 2.0 : Telcos and MVNO's can cash in $	31

Chapter 2 : The rise of Carrier WiFi and Hetnets

This chapter highlights the spectacular growth of Wifi as a viable broadband wireless technology even as Telcos upgrade their networks into 4 G . When debating the choices that mobile operators have for increasing capacity, Wi-Fi, 3G and 4G small cells invariably come up as alternatives or – increasingly – as complementary elements is a multi-RAT strategy. For Wi-Fi to realize its full potential, the industry needs a common set of agreements, standards, implementation practices and interoperability guidelines that outline how operators can easily connect with one another. As Wi-Fi becomes truly secure, seamless and interoperable so new business models and growth opportunities will build on this "next generation" Wi-Fi technology platform. On this multipurpose platform, operators will be able to adopt an offload model to enhance the customer experience or build a roaming business, other players will grow new M2M businesses or create advertising revenues.

2.1	Carrier WiFi in Africa : Key considerations for Mobile Operators	36
2.2	Carrier Wifi Offload : A Strategic Coolio	41
2.3	Wireless Mesh Renaissance : In tandem with 4 G	47

2.4	Hetnets : Carrier WiFi muscles in big time	50
2.5	Carrier WiFi : solution to the African broadband crisis ?	56

Chapter 3 : Planning a Digitised ICT Infrastructure

This chapter delves into the pressing need to establish a digitized ICT infrastructure on a nationwide basis. Besides the direct impacts the real value of the Digital Infrastructure sector lies in its significant impact on the much larger Internet economy and broader digital society. Countries with a strong Digital Infrastructure sector tend to have more individuals and businesses using the Internet. Countries with a strong Digital Infrastructure sector benefit more from the economic impact of the Internet through innovations and productivity gains. The real world is coming online, as smart objects, devices, and machines increase our insight into and control over the physical world. More than just an "Internet of things," it's a new layer of connected intelligence that augments employees, automates processes, and incorporates machines into our lives. For consumers, this provides new levels of empowerment. They are highly informed and can interact and influence the way they experience everything around them.

3.1	Digital Infrastructure : LET'S GO DUTCH	62
3.2	Formulating national ICT Policies : A blueprint for success	68
3.3	The Smart Connected life : Telcos in pole position	74
3.4	National Broadband Policy : A call to action	77

3.5	Smart Meters : Only smart Telcos can play this game	82
3.6	Mobile Signature : Telcos , Bank , Government win for the people	87

Chapter 4 : Gearing up for M&A and Due Diligence

This chapter explores various methods that underpin the financial evaluation of business plans and the savvy thinking behind win-win Mergers and Acquisitions. In reality, not all CAPEX is productive. Investment decisions might prove wrong. Technology and telecom firms are bulking up at the fastest rate in more than a decade, new data show, driven by a continuing explosion of Internet data usage that is hitting everything from handheld gadgets to massive data centers—and the buckling networks that connect them. Behind the billions of dollars are trillions of megabytes of data that are set to be transferred every year on everything from mobile phones and TVs to industrial applications like connected factories and videoconferencing in coming years. Cheap cash from low interest rates is pushing Technology and Telco companies to pursue deals to get better returns, and win scale while they can. And so Banks and Investment firms must understand the technical drivers of a Telco business plan before they dole out capex.

4.1	Spectrum Licenses : Calibrating Valuation and Bid Strategy	92
4.2	Telco M+A in Africa : TDD means TOTAL DUE DILIGENCE	97

4.3	Telco infrastructure sharing : Assessing benefits and overcoming challenges	102
4.4	Broadband business plans : mostly fiction and bamboozle	108
4.5	Broadband economics : Key Insights Reloaded	113
4.6	Carrier Billing is immediate revenue opportunity	119
4.7	Future Track : Augmented Reality is cool and profitable	122

Chapter 5 : Technology innovations drive Network growth and profits

This chapter explores network innovations that will deliver on the nirvana of a connected lifestyle and going Green. Mobile technology is redefining our lives and making it increasingly connected. From health and education to transportation and smarter cities, the proliferation of mobile communication and a connected life is now well established and here to stay. Now, a new wave of connectivity is on the horizon where everyone and everything around us that might benefit from a wireless connection will, in fact, have one. We are about to see connected cars, buildings, medical monitors, TVs, game consoles and a whole range of connected consumer electronics and household appliances. Many of these will be connected wirelessly and intelligently, communicating and interacting with each other, thereby creating the connected life. We will see new applications and services developed backboned on the next wave of social media and messaging platforms to delight the mobile user.

5.1	Energy Saving in BTS : The impact of SON	126
5.2	M2M Reloaded : Crafting strategy to monetise	131
5.3	The Diameter Storm in LTE : GET READY OR ELSE	135
5.4	Inside Track : Carrier Aggregation in LTE	140
5.5	Dumb Pipe Remedy : Transform the packet core to monetise Video / Data	144
5.6	The Mobile VoIP Future is finally here	147
5.7	Telco CEM : Implications in the Digital Era	153
5.8	Getting to grips with Telco API business and Developer Community	158

Chapter 6 : New services to delight the Digital Consumer

The plethora of new technologies and scientific breakthroughs is relentless and is unfolding on many fronts. Almost any advance is billed as a breakthrough, and the list of "next big things" grows ever longer. Yet some technologies do in fact have the potential to disrupt the status quo, alter the way people live and work, rearrange value pools, and lead to entirely new products and services. The technology of the mobile Internet is evolving rapidly, with intuitive interfaces and new formats, including wearable devices. The mobile Internet also has applications across businesses and the public sector, enabling more efficient delivery of many services and creating opportunities to increase workforce productivity. Just as IT creates network effects for users of social media and the mobile Internet, IT-enabled platforms and ecosystems could bring additional value to users of 3D printing or to researchers experimenting with next-generation genomics technology.

6.1	FREE FRANCE : Freaky and Ferocious	161
6.2	Telco Accelerators : Hitting the digital tarmac at top speed	165
6.3	Big growth prospects : Guru Insights on mHealth	171
6.4	Security and IDM : the hidden gold in the smartphone mine	177
6.5	Mobile Money : much more than a gravy train	180
6.6	Monetizing IP MPLS services : Strategic Parameters	184
6.7	Wearables : Worthless Fad or Game Changer	189

Chapter 1 : The Rise of the MVNO

1.1 : Kickstart MVNO's : A look into the WiFi First and Advertising business model

The story of MVNO'S in Africa is not a happy one...infact compared to MVNO's in Europe and Latin America African MVNOs are an ongoing failure for many reasons such as an unfavourable regulatory environment, flawed business models etc. According to Informa Research there are around three million active SIM cards on MVNO networks across the continent, with that number expected to increase to 3.6 million by the end of 2013 and 14 million by the end of 2018. Hardly the stuff of legends !!

We all know that an investment in customer acquisition will tank if the wholesale price is higher than the retail price. Any wholesale rate agreement must have protections against price cutting by the host network so the agreement should incorporate an element of retail minus pricing so that the effective wholesale rate is the lower of the agreed rate or the host's retail rate minus x%. That is why international players like Orange insist that ONLY a formalised MVNO regulatory regime will encourage companies to launch in South Africa and ultimately improve competition.

So what next in Africa ?? Perhaps we might take a cue from what Scratch Wireless and Republic Wireless did in the USA....they have just ushered in a brave new world where Wi-Fi is the default network for free services, and you only use and pay for cellular voice and data if you have no Wi-Fi access. In October 2013, US mobile virtual network operator Scratch Wireless launched a mobile service with an innovative business model that leverages the principle of always preferring Wi-Fi over cellular to allow it to market free voice, data and SMS plans. Scratch Wireless is, at its heart, an innovative MVNO that leverages the growing proliferation of free wifi access , its customised Android smarptphone and its own SMS / VOIP platform to claim unlimited free calling, free data and free SMS over Wi-Fi

Scratch Wireless is a no-contract, pre-pay (pay-as-you-go) MVNO on Sprint's 4G/LTE network that is targeting the price-sensitive, yet tech-savvy, 14-24 year old instant messaging generation. This demographic "uses very little voice" with "average call times of less than four minutes." This, coupled with the fact that the cost of voice termination in the North America is trending towards zero, Scratch Wireless has calculated that it can offer free voice calling (inside North America) while on Wi- Fi. Instead of offering its subscribers Wi-Fi offload as a complement to the more-often-used cellular network, Scratch Wireless has flipped the model on its head and is carving out a space for itself by being the first MVNO to pitch Wi-Fi First.

Scratch Wireless launched with a single smartphone, which comes pre-installed with their own customized version of Android 4.1 Jelly Bean that, according to the company, has been certified by both Sprint and Google. Scratch are keen to point out that the company's "secret sauce" is installed native and is not available via an app. The customization is a smart move, ensuring that the Wi-Fi, VoIP and SMS smarts appear native in the handset. This is important for two reasons. First, by ensuring that the support is native, Scratch has a better chance of controlling quality of service and avoids all the problems that arise with the variability of app performance on different versions of the android OS.

Secondly, the user does not need to do anything to make a free VoIP call to the USA, send an IP SMS or download data over Wi-Fi. All SMS/texts sent over the Wi-Fi network and cellular network are free IP texts sent using the Scratch Wireless SMS platform.

This Scratch features in the OS ensure that all voice calls made while the smartphone is connected to Wi-Fi are in fact VoIP calls that use the Scratch VoIP platform rather than using Sprint's cellular voice platform. Wi-Fi allows operators to improve consumer reach into new mobile devices that emphasize video experience. Offering content for tablets and smartphones via Wi-Fi can reduce the dependency on mobile network operators and guarantee a better customer experience. There is no question that this business model is innovative. But with its high priced cellular voice and data packs, this model really only suits those that are nearly always inside a free Wi-Fi zone or are happy to communicate mainly via SMS and pay a premium for cellular voice ; data services when they need it. In response to Scratch other Wifi First MVNO's are sprouting up.

Working with qualified MVNE players it is possible to reduce capex by leveraging their existing infrastructure (billing , logistics , CRM) to launch an MVNO within weeks. Competitive advantage is achieved by successful MVNOs through effectively leveraging their existing assets to generate customer growth with low customer acquisition costs. As such an MVNO relationship (Wifi First) can deliver revenue from onselling bundled offers (3G Data + Wifi Data + Voice + Mobility). Most SA Mobile operators offer a connection bonus plus 20 to 25 % of on going revenue for their service providers....money for jam !!

Infact Mobile Operators in Africa should not bother with the free wifi game even for data offload : they will have LTE spectrum soon enough , LTE direct and LTE U is on the way and their core competence is managing Macro Cell 3GPP infrastructure...let the ISP's and Municipal councils do their wifi broadband digital divide thing on fibre backbones. However Mobile Operators should facilitate the growth of WiFi MVNO's (like Scratch Wireless) so the consumer gets the best of both worlds: Cellular and WiFi. MVNOs provide MNOs with access to potential customers via non traditional channels such as WISPs and alternative retail stores.Belive it or not there are many ways to make money from WiFi for Telcos and MVNO's even if wifi is dished out FREE....you just have to be a little creative : Through advertising, CSP's have a potential opportunity to:

+ Reduce the price of content and services to end-users
+ Increase the volume of available content and services,
+ Provide value to the advertising community

The idea of offering customers free access to the internet in exchange for watching a short ad from the sponsor allows users to integrate with sponsor brands' social network, increasing their opportunities. For instance, studies have shown that 90 percent of consumers who receive content or services based on their location perceive value in that communication. More important for the advertiser, 50 percent of those recipients act on that information. Mobile advertising can generally be divided into three categories. Landing-page ads are displayed on a site's primary webpage, providing the mobile version of their web portal. Access ads are displayed when the Wi-Fi network presents the sign-in page to subscribers and nonsubscribers. Location-based ads push content and services based on a customer's location.

Telcos and MVNO's can contribute to the development of a differentiated new advertising channel in which users are provided with a portfolio of content and services supported by contextually-relevant advertising. Operators have an opportunity both to provide their own advertising-funded services as well as become an enabler to the advertising community by helping advertisers interact more effectively with their targets (who may or may not be Telco customers).Advertising based on Wi-Fi access promises to change the economics of mobile advertising.

Modern Wi-Fi can provide much more accurate user location (typically within 3 to 5m [9.8 to 16.4 ft] or less) than mobile cellular, allowing for much better targeted advertising. Typically, the use of Wi-Fi is opt-in, meaning that customers are much more receptive to advertising than the alternative spam model. Cisco research reveals that one Wi-Fi service provider is achieving a $24 CPM in a mall in Canada and another is commanding $40 CPM for a mall in Singapore. One operator is even reporting a $350 CPM for Wi-Fi-based advertising in a high-end mall in Dubai. Business modeling by Cisco reveals that Wi-Fi based advertising can conservatively be a multimillion dollar new business for service providers.

*CPM (Mobile advertising is typically receiving $1 to $10 cost per mil (CPM) – the typical pricing metric in the advertising business.}

In addition Location-based services provide a host of analytics and big data that the venues can use to improve their operations or sell to third parties. Aggregating the information available from the Wi-Fi access points provides unique insights into where people go, their common paths, and most visited places. Mall owners, for example, can then use this detailed data to justify higher rents for stores in high-trafficked areas or to measure the impact of signage on customer traffic patterns. Trend analysis and history comparisons of data can show the effectiveness of changes in marketing of store layout. In addition to shopping malls, other large venues such as airports and stadiums can use the data to help improve operations and security.

Retailers can combine the location and user information from the Wi-Fi access points together with customer relationship management (CRM) and customer loyalty data to provide personalized experiences and offers to shoppers at points of purchase in the stores. Equally, retailers are combining location-based services and shopper services to provide additional product information and help customers navigate throughout the store.

Bottom Line : Mobile operators need to accept the terrible truth that they are no longer in the access business and focusing on growing subscriber numbers obliges them to overlook the very opportunities (such as mobile advertising , M2m , Quadplay) and value creation opportunities that Internet brands are rushing to embrace and MVNO's can bring to the party.

MNO's must really must take a proactive role in helping well funded new MVNO's to explore alternative revenue opportunities instead of forcing the tired wholesale retail voice commodity business. WiFi technology allows the creation of new Service Providers that can co exist very peacefully with 3G/4G. And if they do it right they do make money even as they reduce the price of data connectivity and get " have nots " on line.

---------------------------------♠---------------------------------

1.2: MVNO's in Africa : time to rethink and retool the business model

The annual MVNO Africa Summit was well organised and high traffic event : various players in the telecom value chain including bankers jostled around to get a sense of why MVNO's are not taking off in Africa. Deloitte figures showed there are 1,200 MVNOs active globally, 800 of which are in Europe and less than 10 in Africa, a pathetic figure which may belie a massive opportunity for some innovative MVNO business models. "The potential for Africa is at least 500 MVNOs over the next 10 years. What a fantastic opportunity. I don't think there is any other market out there that has the same opportunity " gushed a key note speaker !!

Yours truly was chairing Day 1 of the Event and set the tone by asking the audience why African MVNO's are unsuccessful . I was requested by the Organisers to read a letter from a Ugandan MVNO that is struggling and here are a few excerpts that sortta explains some of the conundrum:

" Tariff regime is not properly regulated with very limited interventions in balancing the market forces. The law of the jungle now prevails i.e. survival of the fittest. Price undercutting is the order of the day, pushing small operators against the wall!"

" In Uganda, 'real' regulation has diminished and almost absent. It's the 'big boys' that have turned themselves regulators through 'arm twisting, control and compromise'. We are seeing emergence of control Cartels! "

" Unfavourable provisions of the partnership and service agreements with the host MNO. We were tied with provision related to capital investments and cost recovery subjecting it into a perennial debt trap. There was reluctance to re-negotiate the un friendly provisions."

" The fear of 'cannibalisation' remained contentious i.e. the Host MNO fear of infringement on its customer base. There was difficulty in harmonising competition between the host MNO and the MVNO. This led to limitations and restrictions in offers for fear of competition thus restricting us to the low return voice with very limited access to the more lucrative data segment.The limited 3G access turned out un competitive compared to offers by other MNOs."

The above sentiments reflect an unfavourable MNO host agreement , brutal price competition , apparent lack of regulatory support and maybe an outdated business model etc which by the way is reality in Africa. So the basic problem may well be that the business of making a margin on the wholesale price is at an end in Africa. The MTR rates are declining even as Operators battle for profit on voice and data by reducing prices. So what can be done ?? Its time for aspiring MVNO's to think about other revenue sources while earning little or nothing on basic voice / data. How about an ad funded business model pioneered by Blyk ??

Blyk was an ad-funded MVNO focused on the 16-24 year old market (although they positioned themselves as a 'media company'). It gifts minutes and texts to customers in exchange for the right to send advertisements to them. Users complete a set of questions about themselves when they sign up, giving Blyk information about their preferences. Advertisers market their products and services via text to Blyk users based on this profiling and Blyk got paid to deliver the advertisement. So, at first glance, Blyk reversed the normal revenue model for operators: it collected money upstream (advertisers) and paid out for delivering services to customers: in reality however Blyk was making money from customers as well (2 sided business model) via :

Termination charges from off-net callers : Operators are both receivers of money from end users (when originating the call) and receivers of money from other operators (when terminating the call). So every time a Blyk user receives a call or text from an off-net customer the originating operator pays Blyk for termination. In turn, Blyk obviously paid some of this termination charge out to its network supplier (Orange) but one can hazard that the company still made some margin on this.

Overage : Typically 16-24 year olds, like the rest of us, have a pre-determined communications budget – "I will spend $x on my phone each month". The fact that Blyk gave users free calls and texts dis not stop users from spending this money. Blyk's users would simply display the same behaviour that every Telco exec is familiar with: increased communications usage as the price reduces (price elasticity). Blyk offered 217 free minutes and 43 texts, so users will be profligate with their communications. They would use this free allowance up and STILL spend at least some of their previous budget.)

Two years after its launch , Blyk became part of France Telecom's Orange network in the U.K. Despite successes in luring major advertisers and producing effective campaigns, Blyk never reached its hoped-for scale. It topped out at 200,000 customers in its first year of existence.A co-founder of Blyk, admitted, "Advertisers want to see scale but we were not rolling out quick enough." Blyk ran more than 2,000 campaigns and attracted quality marketers to its service: Coca-Cola, Colgate, Penguin Books, L'Oreal, Lucozade and the BBC all partnered with Blyk to create cutting-edge mobile marketing aimed at an exclusively young audience. Many people in the industry view Blyk as a pioneer; a trailblazer for mobile marketing whose innovations have set the standard for the industry and will continue to play an important part in its evolution.

Orange's ambition is to grow its media business in order to offer a portfolio of opportunities for consumers and media houses alike. Blyk has proven that sending timely, interesting and relevant messages to young people from brands builds high levels of engagement and elicits high response and action rates. Orange will continue to offer brands the same direct communication but on a far larger scale backboned on Blyk's advertising engine. In addition the latest trend in mobile content advertising is the use of viral marketing. Through specially designed programs users can send recommendations for mobile content they like to their contact lists. A good example for this is the Italian Passa Parola, which has reached a total of 800.000 registered users, by the use of viral marketing alone.So what are the lessons from Blyk MVNO rise and demise ??

Perhaps the most obvious lesson for other operators is that there is value in 2-sided business model where the Telco earns both from upstream and downstream clients : in Blyk's case from advertisers and subscribers ! While Blyk struggled to make a return because of scale (or lack thereof) ,it did enough to show that for operators with large existing (youth) customer bases the ad-funded model could be fruitful. It was not mere marketing fluff that Blyk refers to itself as a media company rather than a MVNO. It showed that Blyk's management considered the advertising community as its primary market and end users as 'members' rather than customers. Indian carrier Aircel teamed with youth media firm Blyk to launch a content and advertising service in November 2010.Blyk still has its eye on the prize: developing the capabilities – in partnership with mobile operators – to be a game-changing engagement media in reach and response !!

Even with its relatively low-tech data acquisition approach, Blyk showed that targeting customers with the right message/product/service/solution really does work. However with your customer base on one side and you are building scale on the other side, the platform will only thrive it not only provides an effective service (identification, authentication, advertising, billing, content delivery, customer care, etc.) and does it more cheaply than can be found elsewhere. Google is winning because advertising is cheap for brands, Microsoft won on Windows partly because the platform, when bundled in with a PC purchase, was negligible. This means that driving costs out of the platform is critical. A strong CRM capability becomes a must-have if they wish to become a platform player like Google.

MVNOs that provide M2M (machine-to-machine)/telemetry services, enable companies and consumers to enjoy new ways to manage and monitor their businesses, lives, and health. Clearly there is enormous potential considering also the hype around MHealth. M2M communications offer a plethora of opportunities to all parties involved, from product manufacturers, to HNOs, MVNOs through to end users. Jasper Wireless was founded in 2004 in California and provides the platform, applications, and design services needed to profitably connect and manage devices worldwide and today partners with prestigious top HNOs such as ATT, Telefónica, Vimpelcom, KPN, Rogers, Telcel and Telstra.

If the MVNO has a strong and INNOVATIVE business case it shouldn't be difficult to prove the real opportunity that can emerge from their partnership with a Host Network Operator. The HNO and MVNO together can create new revenue streams, address new customer segments whilst reducing their marketing costs, increase customer retention by augmented customer satisfaction in the targeted segments and enhance their product offering by creating valuable partnerships with content or VAS providers.

Mobile operators need to accept the terrible truth that they are no longer in the access business and focusing on growing subscriber numbers obliges them to overlook the very opportunities (such as mobile advertising , M2m ,Quadplay) and value creation opportunities that Internet brands are rushing to embrace and MVNO's can bring to the party. MNO's need to take a proactive role in helping well funded new MVNO's to explore alternative revenue opportunities instead of forcing the tired wholesale retail voice commodity business.

And there is no doubt that regulators must play their role to facilitate service-based competition in their markets, while meeting the needs of both existing players and new entrants. Bottom line : there is more than one way to skin a cat !!

---♠---

1.3: MVNO Industry Africa : Learning from the Winners

The MVNO Africa Congress in Cape Town was extremely interesting despite the fact that Mobile virtual network operators in Africa are yet to achieve real success. According to Informa Research there are around three million active SIM cards on MVNO networks across the continent, with that number expected to increase to 3.6 million by the end of 2013 and 14 million by the end of 2018.Hardly encouraging !!

The international speakers really stole the show with their stimulating presentations. Sprint USA '' largest wholesale carrier hosting MVNOs lauded the benefits of partnering with other brands. Sprint hosts 110 of the country's 123 MVNOs and accounts for 8.2 million of its subscribers. Some of the motivations cited for mobile network operators to partner with MVNOs were pre-empting regulatory requirements, selling excess network capacity and addressing additional markets.Infact Sprint has formalised evaluation protocols and raft of support services to on board MVNO's and ensure their success. An enlightened win win relationship based on trust and sincerity is Sprint's " Leit motif "

Vodafone International Services (VIS) surprised us that they have built a 'mobile virtual network operator (MVNO) business in a box' package to offer to clients interested in building a MVNO. VIS, part of the Vodafone Group and based in Cairo, provides a host of services including customer care centres and value added services (VAS) to Vodafone's own networks as well as other players in the telecom and technology industry. They have created a model, backed by our experience as a MNO, of a MVNO business in a box . They have offerings for national and no frills markets, retailers, the corporate segment and niche players.

France Telecom Orange were less enthusiastic about their MVNO prospects in South Africa. They insisted that without favorable regulation for MVNOs they will not launch an MVNO . Any investment in customer acquisition will tank if the wholesale price is higher than the retail price. Any wholesale rate agreement must have protections against price cutting by the host network so the agreement should incorporate an element of retail minus pricing so that the effective wholesale rate is the lower of the agreed rate or the host's retail rate minus x%. This ensures the host network cannot reduce their own customer facing prices below the negotiated prices with the MVNO. Orange is adamant that formalised MVNO regulatory regime will encourage companies to launch in South Africa and ultimately improve competition.

Globally the combination of an increasingly favourable regulatory environment and market maturity has led to a significant growth in MVNOs. Despite the improving environment and the more favourable wholesale terms being offered by networks the list of failed MVNOs continues to grow, demonstrating that creating a viable MVNO remains challenging. The main challenge for an MVNO is no different to any other business and that is to have some form of competitive advantage, a source of differentiation. This is most effectively achieved when the business can leverage existing assets such as existing customers, brand, distribution, content or infrastructure. The greatest challenge is securing a network deal at a wholesale rate which allows the MVNO to create value for shareholders. The chances of negotiating a good deal are significantly enhanced if the underlying business proposition is strong. The chances of successfully securing a network deal are increased if the target host network is selected with care and the MVNO's network architecture is carefully matched to the business needs.

The best case study by far was presented Poste Mobile Italy (division of the Italian Post Office). Mobile communications in Italy is one of the most lucrative, but also the most saturated, marketplaces in the world. In addition, although virtual operators have existed for several years in Europe and the United States, only a few countries have seen MVNOs capture a significant share of the market. Therefore, Poste Italiane saw the success of this new initiative resting on several key factors. The company needed to provide differentiated, innovative services through its mobile offering. It also had to move rapidly through the planning and launch phases in a cost-effective way, putting in place a completely new organization in a very short time frame.

The right technology platform had to be created that could enable the company's 13,800 existing postal offices to act as the PosteMobile retail network. At the same time, the integration of existing postal offices went beyond technology—it also required significant organizational change management skills and a large training effort. Post Mobile worked with an MVNE who provided the support for the core elements needed to launch and operate the new MVNO.These included:

- Business support system (BSS) functions such as customer relationship management, billing, dealer portal and data warehouse
- Network elements, including IN Service Control Point and GGSN
- Service delivery platform, including a mobile portal to enable SIM-based value added services

PosteMobile's market leading services allow customers to conduct a variety of financial and communications activities easily, securely and inexpensively. Customers can check their PosteMobile accounts, make money transfers and pay bills through the easy to use SIM menu. They can refill their accounts easily, and can monitor the movement of funds in both the PosteMobile prepaid card and Poste Italiane accounts. PosteMobile and Poste Italiane have achieved important business results that are propelling them toward high performance. In the first month of operation as a startup company, PosteMobile attracted 100,000 subscribers and now has over 4 million subscribers—making it the largest virtual mobile network operator in Italy.

Post Mobile Italia is a classic : leveraging their existing customer base and loyalty they built their own billing system and combined mobile payments and VAS into a devastating MVNO success. Poste Italiane has revolutionized the offerings of the traditional postal industry and set new terms of competition. The company can now provide customers with new, highly attractive mobile services by integrating and making its traditional offerings (national post office network, bank and certification authority) accessible through any mobile handset incl NFC believe it or not ! Viva l'Italia !! A little creative thinking does reap financial dividends.

---♠---

1.4: MVNO Success ? Its all about getting Maths right

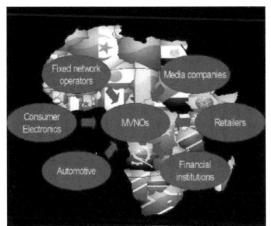

One should expect a powerful growth phase in MVNO expansion in Africa since there are still many market segments untapped by the mobile network operators that are being addressed by MVNO's. These include the ethnic markets, data-only connectivity, and community-led and retailer-owned MVNOs – and there is also a strong push from mobile operators wanting to increase their share of mobile wholesale revenues.

Almost all analysts agree that the opportunity for mobile operators to take advantage of MVNOs outweighs their competitive threat. Moreover, the competitive threat argument is questionable considering prices would continue to fall even without the presence of MVNOs.Ultimately, there will be an increasing need for mobile operators to fill their networks (e.g. 3G or 4G), regulators will demand further roaming and interconnection reductions, mobile-only operators will use lower prices to advance fixed-mobile substitution, and new companies with disruptive technologies (like VoIP) will compete by offering even cheaper voice packages.

Analysis indicates that non-telecoms based MVNOs are able to target profitable market niches and attract more customers by taking advantage of their recognizable brands and well tailored mobile offerings. These players are able to explore the market potential more comprehensively than companies with a telecoms background (e.g. fixed operators entering the mobile market or CATV operators turning to quadruple-play). What opens exciting possibilities for MVNOs offering services beyond discount voice and SMS is also the popularization of 3G+ networks enabling new applications and services, thus leading to further specialization. For example :

• Data-only MVNOs bank on mobile data services to up-sell and cross-sell core products for instance laptops, e-book readers or services for instance telematics or ICT solutions in general — and thus add value to product or services that they already offer within their portfolio.

HP, for instance, launched its data-only MVNO service already in 2009 to improve market share in Japan where it faced strong competition from local manufacturers. The same logic was followed by Amazon that bundled mobile data services to its Kindle to protect its market from Apple and other tablet manufacturers.

Retailers can help MNOs increase loyalty and reduce churn by using their customer-loyalty experience to offer discount and loyalty cards and to offer leasing programs for device purchases. As the VAS market develops, the retail MVNO is also in a good position to develop mobile commerce. The m-commerce and m-banking opportunity, alongside the retail network, is at the core of the potential value proposition of the financial MVNO as well . Furthermore, retailers will naturally target the mass market, which for MNOs raises the risk of market-share cannibalization, especially in the prepaid-dominated African markets.

The financial MVNO faces the same challenges as the retailer MVNO in terms of negotiations with MNOs, given its natural focus on the mass market. In addition, operators are increasingly eager to play a role in the m-commerce and m-banking markets. The way regulation develops in this area will largely determine the development of this MVNO model. There is certainly scope in the region for prospective ethnic MVNOs, targeting communities of migrants and nationals with family or friends living abroad to provide financial services.

Postal Organizations in Africa are well positioned to launch their MVNO's now. As a virtual operator, Poste Italiane extended access to its existing postal and financial products via mobile phones, while leveraging wireless technologies to offer innovative, revenue-generating services. In the first month of operation as a startup company, PosteMobile attracted 100,000 subscribers and captured 7,000 activation orders per day at the peak of this growth. Within two months of the launch date, PosteMobile became the leading Italian MVNO in terms of customer base.

The business plan plays a number of essential roles in the planning and creation of a successful MVNO. The business plan covers all aspects of the company launch plan including market assessment, funding requirements, financial analysis, market segmentation and product differentiation. Also included is go-to-market plan, distribution and replenishment plans, comparison of MVNO's and more.

An MVNO is no different to any other business and must have a source of (sustainable) competitive advantage if it is to create value for investors. Competitive advantage is achieved by effectively leveraging their existing assets to generate customer growth with low customer acquisition costs. It is this leverage that provides the basis for a good business "story." MVNOs typically seek to leverage the following assets:

- Existing customers – it is easier to sell a new service to existing customers than it is to win entirely new customers (e.g. Tesco)
- Brand – to be successful the leveraged brand must drive the purchase of mobile telephony (e.g. Virgin)
- Distribution – existing channels to market will help reduce the cost of customer acquisition (e.g. Tesco, Aldi)
- Content – for some, mobile provides simply another media for the distribution of existing content (e.g. Jim Mobile)
- Convergence – bundling of multiple communication services is increasingly common and has been shown to increases customer loyalty (e.g. Tele2, BT leveraging customers and fixed assets)

Along with the evolution of MVNOs and their growing needs, a market of mobile virtual network enablers (MVNEs) has emerged. MVNEs provide MVNOs with all necessary back-office operations and IT platforms allowing them to concentrate on the core of their mobile business – developing new tariffs and services and taking care of customer acquisition and retention. The MVNO paradigm provides companies with access to revenue-generating goods and services and transfers their specific experience to the wireless markets.

On the other hand, mobile operators acquire access to new markets and additionally, they gain access to specialized content and services. Sharing some business processes will certainly result in their overall performance increasing. With vast amounts of unused bandwidth to share and the inability to provide proper services to all market segments and niches by traditional operators, the MVNO is a classical win-win situation with greater economies of scale and more value added for end-customers.

Becoming an MVNO requires careful preparation and recognising at the outset that mobile operators will only entertain partnerships with MVNOs' that:
- Target niche segments that the mobile operator is unlikely to serve and have good scale or customer value potential;

- Offer a unique service proposition that minimises the likelihood of cannibalising that operator's existing customer base;
- Have the brand reputation, channels to market and execution capabilities in their current market;
- Have the financial resources to support the project through to break-even.

In practice this means only approaching the mobile operator when fully ready with a compelling, clearly differentiated and commercially attractive proposition and with at least $4 million in secured funding. Most wannabe MVNO's fail because they underestimate the initial capex. If you don't have money don't bother to play the MVNO game !!

1.5: Enterprise 2.0 : Telcos and MVNO's can cash in

In its most basic form, Enterprise 2.0 is about communication. When information is free, people can get more feedback and input (collaborate), react more quickly (agility), and make better decisions. This is the opportunity inherent in Enterprise 2.0: a more efficient, productive and intelligent workforce.

The current crop of Collaborative solutions focus around unified communications (instant messaging, web conferencing and VoIP for example), working in teams, sharing documentation and knowledge, working with (self-service) portals and working with social collaboration tools.

Organizations are beginning to take advantage of social collaboration aspects like communities, blogging and wikis to connect with external parties like partners, customers and local government. A survey done by McKinsey & Company showed that companies that benefit most from B2C/B2B collaboration are:

- Networked organizations;
- Business to business organizations;
- Big companies (> $1 billion revenue);
- International companies;
- Decentralized organizations.

According to Industry experts there are three fundamental ingredients to be successful with E20/Social Business (or any major corporate initiative): Adequate resources/budget, organizational commitment, and a business problem to solve. Missing any of these greatly slows down and/or blunts the outcome of the effort. The top challenge is culture change. You can drop social technology into any organization, but you can't suddenly expect that employees will adopt the way that social media works or that business processes or traditions will automatically change.

Social is a new way of operating (observable work, openly participative processes, co-creation) and this requires conscious effort to change our thinking and the way we function. Other top challenges include enterprise apps with overlapping features (e-mail, CMS/DMS, IM, unified communication, enterprise microblogs, customer forums, CRM, etc.), underinvestment in community management, and lack of executive understanding or buy-in.

Web 2.0 is the term for web-based tools and services that allow for – and even improve with – user participation. The most well-known examples of this technology are found in sites like YouTube , Facebook , Wikipedia and Amazon, where users to find and connect to what they are looking. Social media tools like blogs and microblogs (Twitter) opened up the world of media and publishing to anyone with an internet connection – or a smartphone.

Social network tools help staff find the right individual or group of people.Tagging and rating provide a straightforward way to find content and make judgments about what to look at. Blogs and wikis are natural collaboration and communication platforms.Giving employees the freedom to speak their mind and voice ideas is required for there to be a harnessing of collective intelligence

One of the biggest car lease firms worldwide had a clear vision on the use of social collaboration, both internally and B2C/B2B. In this vision it outlines the many strategies it intends to use for social collaboration.

* Launching an employee community
* Engaging in social recruitment
* Social software enabled car remarketing
* Launching fleet management communities
* Social software enabled car quotations
* Launching driver and supplier community
* Launching a supplier community
* Conducting online reputation management

A Global management consulting, technology services and outsourcing company implemented off the shelf social media platforms technology to introduce knowledge sharing communities and social networks. Blogs and wikis function as collaboration tools, and as such, they have uses mainly in sharing "unstructured" information associated with ad hoc or ongoing projects and processes, but not for "structured informational" retrieval.

However, Shell has started converting its official documentation to wikis, because this enables that company to make documentation updates available in real time and allows non-editors to contribute to the documentation. In this process Shell restructures the paper documents to a set of on-line wiki pages. Their key challenge was to get key stakeholders aware of how social computing can solve business problems and be integrated into business processes. The business case was based on the following metrics:

- Finding people and identifying experts;
- Finding information;
- Reducing the need for travel;
- Speed up the decision making process.

Current Telco mainstream offerings to the Enterprise market are based around capacity and hosted services, sometimes complemented by IT outsourcing projects. Mass-market consumers and Enterprise customers alike are increasingly demanding rich, portable, personalised, access and device-independent services from their Telco Service Providers.Telco 2.0 and Web 2.0 components creates more value to the Enterprise . For example Telco resources can be embedded with the Enterprise applications to identify the real time location and distribution of a service engineer's customers (using Google Maps and Location feeds) to view the geography of the area covered.

In the UK, BT (British Telecom) has become one of the country's strongest proponents of enterprise 2.0. The company has introduced a raft of social media tools, including a huge Wikipedia-style database called BTpedia, a central blogging tool, a podcasting tool, project collaboration software and enterprise social networking.

With SDP / IMS platforms Telco customers can become part of the social networking phenomena by complementing these content based sites with telecom capabilities such as anonymous calling (whisper calls) and real time updates showing the physical location of friends and contacts within the community group. Whilst Telecom Web Services standard will revolutionise Telco service offerings to both the Large Enterprise and the SMB market, it is important not to overlook the benefits of being able to offer fully hosted services.

These include "Virtual PBXs" and "Virtual Contact Centres", plus a suite of complementary services to customer's own installed platforms such as "Mobile PBX Extensions", "Multi-line" services for handsets and of course "Voice Call Continuity" to provide Enterprise roaming in WiFi hotspots.The availability of Enterprise 2.0 tool combined with high speed networks smartphones and cloud computing will unlock fresh new revenue streams for agile Telcos and CSP's in the Enterprise / SMB markets.

We are on the cusp of a management revolution that is likely to be as profound and unsettling as the one that gave birth to the modern industrial age. Driven by the emergence of powerful new collaborative technologies, this transformation will radically reshape the nature of work, the boundaries of the enterprise, and the responsibilites of business leaders.

Chapter 2 : The rise of Carrier WiFi and Hetnets

> 2.1: Carrier WiFi in Africa : Key considerations for Mobile Operators

While many Carriers will continue to pursue technological advancements to handle demand, they maybe reticent about Carrier Wifi in their RAN portfolio because of high cost of integration , OaM and apparent lack of business case . At the same time, long-term spectrum availability, spectrum efficiency , LTE Femto cells and continued backhaul improvements are likely to be a key focus to assure continued mobile broadband momentum.From an MNO perspective, a major concern of Wi-Fi spectrum is the fact that it is shared with many other users and operators, usually unknown, and the traffic is unmanaged or coordinated. In busy areas, this can lead to congestion in public areas, manifesting itself as slower data rates or no connectivity.

A major problem for mobile operators is that the current trend for multi-mode networking (i.e. combination of WiFi and 3G access) limits the ability of operators to provide VAS services and/or capture 2-sided business model revenues, since so much activity is off-network and outside of the operator's control plane. As MNOs will face traffic congestion challenges over the next 5 years, they will need to make technology choices regarding capacity enhancement in the near future. Opting for small cells based on mature cellular based standards may be seen in some markets as the technically superior and least risky option. At the current time, this is the route favoured by many visionary MNOs outside Africa .

To make matters worse slow regulation processes have negatively impacted Africa's broadband deployment. The delay in local loop unbundling has either discouraged potential players in the fixed broadband market or forced them to invest in their own infrastructure, making the deployment limited and targeted at more profitable areas. Spectrum resources could derail plans for South African mobile operators. Analysts strongly believe that the quality and breadth of mobile content is key to operators' outperformance in the 3G/4G market, but note that as operators encourage greater usage of data intensive services such as music and video steaming, their limited spectrum resources will fall under increasing strain

Mobile operators have been able to refarm existing 2G spectrum in order to launch LTE services and consolidation in the telecoms sector is also expected to help address spectrum scarcity. While Vodacom's planned acquisition of Neotel is primarily related to its extensive fibre coverage, it will also give the mobile operator access to Neotel's large chunk of spectrum in the 3.5GHz band. However, with operators now under even greater pressure to boost data revenues and broadening access to data intensive services. Most Analysts believe that any further delays in spectrum allocation could threaten their ability to maintain quality of service standard on their networks.

There are too many players in Africa, so there is a strong likelihood and need for consolidation. Under resourced telcos should be obliged to use their spectrum or return it to the Regulator instead of hoarding : this combined with sluggish Regulators is the bane of so called spectrum dearth Africa !! Despite the clear economic rationale, there may still be some obstacles to consolidation, such as the unrealistic ambitions of owners and senior management or an investor reluctance to take a financial write down. There is a nagging perception that Telco / Wireless operator assets (spectrum , tower , fibre , IT Core , brand equity , subscriber revenues etc) in Africa are overstated and overvalued. So it should not be acquisitions for the sake of acquisitions. Without scale economies future success is at risk.

According to GSMA where necessary, mobile network operators will tend to use small LTE cells to relieve congestion in traffic hotspots rather than relying on Wi-Fi. The prospects for Wi-Fi delivering significant capacity relief in areas of the cellular network facing congestion are limited. On the contrary, Wi-Fi and cellular traffic are expected to grow in parallel and rapidly, offering complementary capabilities.For MNOs, the ability to monitor traffic levels in the band, steer traffic between bands and cells and implement policy rules are key requirements. Wi-Fi does not provide this level of visibility and control, except for Wi-Fi deployed by the MNO itself. While there exist functional and performance short-falls with the use of Wi-Fi for small cells, they should not prevent its use in some scenarios. It could be challenging for example for operators with limited 3G or 4G spectrum allocations or those unable to deploy small cells and backhaul on their own. In this case, a Wi-Fi based solution could be justified.

Perhaps the most transparent example of usage patterns on operator-deployed Wi-Fi networks comes from the world's largest mobile market, where market leader China Mobile has regularly reported the distribution of traffic across the world's largest public Wi-Fi network by number of hotspots, which at end-June 2012 stood at a total of 2.83 million access points across hundreds of thousands of hotspot locations. The data shared by China Mobile highlights an alarming statistic which is the discrepancy in the amount of revenue generated per megabyte by China Mobile on traffic that flows across its cellular and Wi-Fi networks. China Mobile, it generates US$0.0367 for every megabyte transmitted on its cellular networks, but just US$0.0004 on Wi-Fi. China Mobile would generate an additional $ 1.4 billion in annualised revenues for each 5% of data traffic that was monetised via its cellular network instead of over WiFi.

As the telecom ecosystem expands, beyond the unending need for broadband access to content and speed, new business models should be explored among both traditional and new players, which could open up markets and change the landscape. Deployed effectively, Carrier WiFi solutions can enhance consumers' personal and professional lives in realms including education, health care, automotive, hospitality and beyond. Telcos must focus on " exponential technology innovation " built on proactive, thoughtful planning of the ecosystem the business operates in. Ongoing, accelerating advancements in computing power, storage and bandwidth in smartphones are providing the platform for disruptive innovations, an impact that is amplified when technologies coalesce into open platforms and ecosystems.

There is no doubt that the smartphone will be the first departure point for users seeking to gain access to digital services such as e-commerce, information and other services.New devices with wifi chipsets will continue to arrive with varied price points and different form factors in smartphones and tablets. Mobile capabilities are being extended into completely new devices, including wearable technology such as glasses, smart watches and fitness/health devices etc. The ecosystem evolving around online – including mobile – substantially broadens the availability and distribution options for content. Smartphones and tablets are directly contributing to this shift and increasingly drive where and how content is delivered. While the availability of low cost wifi enabled smartphones will boost internet connectivity in the low income prepaid user segments the current pricing for " Out of Bundle " data will limit the actual network traffic and alienate MBB subscribers.

New entrants will come into the market from adjacent sectors – broadcast media and even advertising groups are viewing telcos as a ready-made channel to market. The rise in online video consumption and faster access technologies (Fttx , 4G) has sparked a slew of VOD initiatives in the last few weeks enabling wireless operators and digital content providers to bypass traditional satellite broadcasters. The rising trend in Internet advertising might indicate collaborative partnerships to potentially subsidize low income segments The aggregation of content over a larger subscriber base or the aggregation of advertising agencies across a larger base of broadcast platforms could result in respectively lower direct costs or higher revenues.

The evolution of the Wifi Standard is opening up new application opportunities in the IoT and M2M arena. The availability of Cloud as the orchestrator of various Wifi capabilities would suggest an entirely virtualised technology architecture to deliver and manage connectivity. Cloud is rapidly becoming important for many telcos; small businesses as well as public administrations require some form of cloud service. In this context, M&A can serve multiple purposes, either with regard to accelerating market timing, shaping its structure, or acquiring specific skills and technologies (e.g. lightweight, scalable, low-cost infrastructure.

Despite the rosy WiFi technology scenarios Mobile operators need to respond to a number of important questions before they incorporate Carrier WiFi into their RAN portfolio and Small Cell Strategies . What is the impact of a seemingly unstoppable transition to free-to-end-user Wi-Fi in public locations on users' perceived value of Internet access on the go? What impact is user dependency on Wi-Fi having on their willingness to pay for bigger data plans or to deliberately avoid incurring (lucrative) overage charges ? How does Carrier WiFi fit into small strategy with the arrival of LTE ? Should WiFi rollout be left to ISP , FTTx players and Municipal / Govt free wifi projects if the MNO's get sufficient 4G Spectrum ? does it make economic sense and under which circumstances should MNO's deploy WiFi for data offload ? Is the really worth the MNO's effort to deploy WiFi with the arrival of LTE Direct ?

As the China Mobile case has served to highlight, there is a major monetization gulf between the ability of mobile operators to generate returns from managed public Wi-Fi traffic compared with when it flows over their cellular networks.

---------------------------------------♠---------------------------------------

2.2: Carrier Wifi Offload : A Strategic Coolio

The first Wifi Offload Summit in Africa was a long overdue but welcome event indeed .The cross industry speakers at the Event (including yours truly) waxed lyric about the technology benefits..and its titanic revenue potential (did someone say a billion dollars in 5 years)..hmmm... wishful thinking is always prevalent at the peak of inflated expectations in the hype cycle !!! But we all know that only a market sizing study to assess the real revenue potential into both the consumer and industrial Wifi segments can calibrate the spin our technology roadmaps and marketing plans.

African Operators have been rather cautious about wifi in their RAN portfolio for many reasons : they thought wifi was not carrier grade since its was not 3GPP " certified " , their circuit switched GSM generated huge revenues from voice / sms without capacity constraints , there wasn't enough submarine bandwidth , the price of smartphones was too high and so on . It is only now with the arrival of low cost wifi enabled smartphones and huge OTT data surge on packet switched cores that African telcos are facing network congestion and revenue decline (voice / sms) ....and unfortunately the subscribers higher data consumption does not linearly translate into higher revenue.

So what is Carrier Wifi ?? In carrier Wi-Fi, a mobile operator or an ISP, owns and operates the Wi-Fi infrastructure, manages access from users, and shares the infrastructure with roaming partners. Mobile operators may choose to integrate Wi-Fi within the cellular network, both within the RAN by co-locating cellular and Wi-Fi small cells, and in the core network by integrating authentication, subscriber management, billing, policy, and traffic management. To integrate Wi-Fi in the cellular network, trusted (i.e., run by the operator or its partners) Wi-Fi APs connect to a Wi-Fi gateway that interfaces the mobile core. 3GPP efforts in defining such interfaces and additional functionality to support Wi-Fi integration have been crucial to provide a framework that integrates cellular and Wi-Fi access in the mobile core.

China Mobile is the poster boy of Carrier Wifi : their aggressive Wi-Fi strategy that has led to the installation of 2.83 million APs in its hotspot footprint. In the first quarter of 2013, Wi-Fi traffic on China Mobile's network accounted for 73% of overall traffic, up from 50% in 2010. In 2012 Free France announced that it has opened up its 4 million-hotspot community Wi-Fi network to its smartphone customers, creating the world's largest carrier-run mobile data offload network. By leaning heavily on Wi-Fi, Free can offload enormous amounts of traffic that would normally traverse HSPA+ networks, where capacity is scarce and bandwidth expensive to deliver. Ongoing industry initiatives like NGH and Hotspot 2.0, and vendor solutions are unlocking the opportunities

(reduce network congestion , incremental revenue etc) for Carrier Wi-Fi. Yes : Carrier Wifi as with LTE is all about SCALE to leverage the network externality !!

In addition 3GPP and Wi-Fi Alliance, are developing a standard architecture to enable Wi-Fi interworking and roaming with access to voice, video and multimedia services in the carrier's 3G or 4G core network. The iWLAN architecture, standardized by the 3GPP, promotes secure connections between the packet core network and Wi-Fi networks. The iWLAN architecture designates a secure connection to the carrier's core network via an IPSec tunnel, which extends from the UE, through the wireless access gateway (WAG), the GGSN, and from there to the core network or the internet. Carriers may provision, authenticate and authorize traffic for their carrier services on Wi-Fi networks, just as they do for their 3G networks, and they can enforce policy on these services. 3GPP IWLAN is future-oriented and as such is specified only for 3G RANs and handsets and requires SIP and core IMS equipment investments and depends on an embryonic network function—the voice continuity server (VCC) to enable roaming from and to WiFi and GSM access networks.

When debating the choices that mobile operators have for increasing capacity, Wi-Fi, 3G and 4G small cells invariably come up as alternatives or – increasingly – as complementary elements is a multi-RAT strategy. Both Wi-Fi and cellular small cells act as an under-layer to the macro network and bring the mobile network closer to the subscribers – down to street level or indoors. Wi-Fi and cellular small cells address the need for additional capacity in subtly different ways, and this makes them complementary, not mutually exclusive. In some environments – e.g., indoor locations with high traffic loads – Wi-Fi may be better suited. In others – e.g., a park or open urban space – 4G small cells may work better. While planning for a 3G and/or 4G small-cell deployment with integrated Wi-Fi, the differences between cellular and Wi-Fi interfaces are crucial.

The combination of 3G/4G with Wi-Fi gives the mobile operator a good comfort zone. At all times the operator can count on a capacity level from the cellular network that is manageable. Typically the additional capacity from Wi-Fi ensures a higher QoE and hence subscriber satisfaction, but this cannot be guaranteed because the operator does not control the spectrum. According to TCO sensitivity analysis from Senza Fil it appears that Wi-Fi and 4G is the winning combination, because the increase in per-bit TCO is less than the increase in capacity – although the savings over using Wi-Fi alone is small when measured against our benchmark of 4G-only per-bit TCO (42% for the Wi-Fi and 4G combination, versus 43% for Wi-Fi only). More notably, that 42% means operators can have 4G plus Wi-Fi for less than half the per-bit TCO of 4G alone.

To evaluate the potential Wi-Fi hotspot business models and revenue opportunities, carriers must consider their target audiences and how those potential customers use the Wi-Fi hotspot network. At any given Wi-Fi hotspot, a diverse mix of customers may have widely different needs, expectations and device or payment capabilities. Overall, there is no one-size-fits-all business model, and the network architecture and level of integration with existing services will reflect the unique business goals of each carrier in each different market. Analysts believe that the strategic architecture to differentiated Wi-Fi services for customer acquisition, retention and monetization necessitates a careful balancing of the following parameters :

+ Understand capacity and revenue optimization : enable dynamic offload and onload between mobile and Wi-Fi networks utilizing device and network information on subscriber plans, billing cycle, data usage, personal Quality of Experience and network conditions.

+ Enhance customer experience : automatically generating policies based on real-time and predictive Wi-Fi Quality of Service to balance the load between Wi-Fi and mobile networks and improve the subscriber experience.

+ Integrate roaming plans : using ANDSF discovery information and policies for partner Wi-Fi network access and authentication to seamlessly offload mobile subscribers to partner Wi-Fi networks based on subscriber's data plans and entitlements.

+ Leverage context-aware promotions : capturing and using granular data on connections and user behavior – such as indoor location, dwell time, user travel and subscriber preferences – to deliver targeted offers and advertising directly to the subscriber device.

The humble Wifi radio never ceases to amaze : it is has crossed over into the Internet of Things (IoT) and machine-to-machine (M2M) worlds. The arrival of low power , standards based Wi-Fi chips heralds a new wave of Wifi applications and revenue for savvy Telcos. Configurable Low Power WiFi is just a fancy term of Wake on Wireless Lan (WoW), which allows the system to be turned on or awakened by a network message. Relative to Zigbee/802.15.4, low power Wi-Fi takes advantage of the benefits conferred by the well established IP and Wi-Fi protocols .

Low power Wi-Fi semiconductor solutions are starting to appear on the market addressing applications such as thermostats, water heaters, HVAC systems, blood pressure, glucose monitoring and other healthcare devices, asset or people tracking tags, etc. Wireless Sensor Networks (WSN) market will grow to $1.8 billion by 2024 according to research reports. Wireless Sensor Networks will eventually enable the automatic monitoring of forest fires, avalanches, hurricanes, failure of country wide utility equipment, traffic, hospitals and much more over wide areas, something previously impossible.

But here is a reality check for African Telcos : successful Carrier Wifi works best with fibre optic backhaul. And terrestrial fibre optic in Africa is still a SLOW work in process. The burning questions for African telcos : when to offload and how to do it ?? What about the arrival of LTE u ? Is it better to have technology consistency in the RAN portfolio ? Are sensor applications in industry verticals a more profitable market to pursue ? Is the terrain best covered by macro cells ? Is a standalone wifi network to offer different applications and services a viable option ? what about availability of locations to set up Hot Spots ? do i really need to offload or can i use cheaper technology mechanisms (like policy control , network optimization etc) to reduce congestion and enhance the QoE etc ? .Can we create an aggregated Wifi Cloud on existing wifi assets in Africa ? Many questions need to be asked and answered in a sober manner before you embark on the Carrier Wifi journey.

Bottom line:don't jump into Carrier Wifi without a thorough Opportunity / Risk / Reward analysis because monetising Carrier Wifi Technology investments is much more complex than it seems.In every case a solid business case must be constructed based on all technical and commercial variables that influence the ROI through the Hype Cycle into the plateau of productivity.And...please make sure you have sufficient CAPEX before you build out a Carrier Wifi network !!

---♠---

2.3: Wireless Mesh Renaissance : In tandem with 4G

Wi-Fi has already delivered secure, high-speed Internet access from thousands of "hot spots"— restaurants, libraries, schools, bus terminals, airports or other public places via millions of APs. Wi-Fi evolved from IEEE 802.11 (3Mbit/sec.) to 802.11b (11Mbit/sec.) to 802.11g/a (54Mbit/sec.) and now at 802.11n (300 Mbit.sec). Although the transmission rate has risen dramatically, the range of is still limited. Meshing is an extension of Wi- Fi that takes advantage of high data rates by having every wireless device act as a router/repeater. This enables very long links—even several miles—between users and APs by "hopping" through a series of short links. While the range of Wi-Fi is limited per link, meshing lets users string these links together to cover a virtually unlimited range. The coverage area of the radio nodes working as a single network is sometimes called a mesh cloud.

Meshing is based on military-funded research designed to meet the requirements for battlefield communications, a major advancement from the Wi-Fi networking technology. Devices enabled with Wi-Fi chips can send and receive information anywhere within the range of an access point (AP). The mesh principle is similar to the way packets travel around the wired Internet— data will hop from one device to another until it reaches its destination. Dynamic routing algorithms implemented in each device allow this to happen. Meshing creates a self-forming, self-balancing network that makes deployment significantly less costly, faster and more robust. The technique is less expensive both in installation and ongoing monthly costs because it reduces the total number of backhaul links needed. Users can cover large areas from a single Wi-Fi AP/backhaul combination by using wireless routers to extend coverage over large distances for much less cost than putting up multiple APs.

Because a meshed network is self-forming, network providers needn't spend a lot of time and money re-engineering the network. If more coverage is needed, they can drop in a wireless router to automatically extend coverage. Providers can cover areas that don't have backhaul available such as a large park, beach or body of water—by deploying routers. Meshed networks are also self-healing and self-balancing, so traffic congestion or failures at a particular access point can be resolved automatically. For example, if an AP at a popular Coffee shop becomes overloaded with users, meshing automatically causes some of the users to "hop" over to nearby APs. Network utilization, an important parameter for network operators, increases dramatically because of this self-balancing act. From the user's standpoint, there's higher performance and the potential for lower service costs.

An IBM's research study called the "Citywide Broadband Study," examines how these autonomic networks could change the way communities work and play. A number of municipalities in the USA are learning that practical applications of mesh networking go well beyond the emergency-response example cited above. Each device in a city— traffic signals; message signs; public transit assets such as buses, electronic information kiosks and video cameras—would become part of a grid of wireless devices that communicate with one another and transfer information throughout the network.

The city of the future a.k.a SMART CITY will deploy a large-scale mesh network to keep their workers and first-responders productive, no matter where they are in the community. A single infrastructure that supports many different municipal applications and departments ultimately reduces networking costs and simplifies operations. With pervasive Wi-Fi, court officers, building inspectors, transit workers, social services workers, and other city employees can perform their duties effectively while in the field. Wi-Fi hot zones also support business development and are a convenience for the general public. Cities can install telemetry and smart grid services using mesh networks to support automated traffic control, smart utility meters, and smart parking meters. Mesh networks also enable sensors used for earthquake, tsunami and gas detection, among others.

Wireless mesh networks are ideal to connect industrial operations and sites such as oil and gas fields, mining and construction areas, which are difficult to network because of their geography. With pervasive Wi-Fi, field workers communicate easily and have access to key applications. Mesh enabled IP video surveillance and access control also protect the organization's field operations and crew.

Even cutting edge Mobile Operators are using carrier grade Wireless Mesh as part of the Data Offload and incremental revenue strategy. In 2010 Mobily Saud Arabia (who launched the first TD LTE commercial network in the world) noticed a massive increase in data usage, fuelled by offering unlimited data subscriptions. Against this backdrop of increasing data usage, Mobily saw Wi-Fi as an efficient way to reduce the cellular capex investment in broadband infrastructure needed to match this spike in data usage. Today Mobily has around 350-400 public hotspots with each Hotspot comprising of multiple Wi-Fi Access Points covering multiple business verticals including cafes, hotels, hospitals, outdoor, and some challenging venues such as stadiums and Holy Pilgrimage areas.

Mobily's business model is predicated on a hotspot portal based Wi-Fi virtual network with multiple service monetization models for both Mobily and non Mobily subscribers PLUS a cellular-to-Wi-Fi offloading virtual network, offering seamless and secure user experience with the use of EAP-SIM protocol and WISPr clients. Mobily intends to "offload" at least 20% of current mobile broadband traffic onto Wi-Fi networks and is designing the Wi-Fi network to meet this key performance objective . The Hotspot 2.0 standard will open the door for inbound roamers to connect seamlessly and securely to Mobily's Wi-Fi network while their usage is being charged back to their home operator.

---------------------------------------♠---------------------------------------

2.4: Hetnets : Carrier WiFi muscles in big time

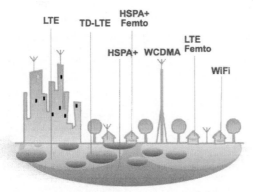

The shift from wired to wireless is well underway and as mobile demand continues to skyrocket, capacity continues to shrink. According to Cisco, global mobile network data grew by 70% in 2012, amounting to 885 petabytes per month, up from 520 petabytes per month at the end of 2011. Total data volume in 2012 was a staggering 12 times more than the size of the entire global Internet in the year 2000. By 2017, monthly global traffic is expected to reach 11.2 exabytes, which is 1144% higher than in 2012, an average growth rate of 66% per year.

A recent survey revealed that tier-one mobile network operators expect 22 percent of all additional data capacity added during 2014 to come from Wi-Fi offload, according to the Wireless Broadband Alliance.By 2018, Wi-Fi offload is predicted to contribute 20 percent of additional mobile data capacity plus a further 21 percent will come from small cells with integrated Wi-Fi.These findings demonstrate how important data offload is to mobile network operators, accounting for an average of 20 percent of data traffic, up to 80 percent in densely populated areas such as transport hubs and cafes. Within homes and businesses offload levels are 50 to 60 percent.Telstra will spend more than $100 million to build a wi-fi network by next year involving around 8000 Telstra-built hot spots and a further 1.9 million wi-fi access points provided by its customers nationally.

The Cisco Visual Networking Index report that a laptop can consume up to 450 times more bandwidth than a simple phone and 30 times more than a smartphone. The estimated capacity consumption of an iPhone is estimated to be 30 times that of a simple phone.Even assuming some reduction in the rate of traffic growth over the next few years (expected by most data traffic forecasts), mobile operators still need to increase network capacity quickly and cost effectively, without relying solely on new spectrum availability. In this context, densification of the network infrastructure with both small cells and Wi-Fi is necessary for operators to keep up with demand without breaking the bank.

With traditional network design strategies, mobile service providers essentially have three primary capacity expansion tools, including:

- Increase macro cell site density: Each cell split would require regulatory approval, new sites and civil work.

- Technology upgrade to OFDMA-based 4G technologies WiMAX and LTE: To achieve a 3 to 4x increase in capacity.

- Expand radio spectrum resources: Acquiring new spectrum can be expensive, limited by availability and subject to government regulatory timelines. In each case the backhaul transmission network may also need to be upgraded accordingly.

The HETNET can consist of different cell scales which range from macro to micro, pico and even femtocells, potentially sharing the same spectrum. Nodes can deploy different access technologies over both licensed and unlicensed bands. Co-channel femtocells can provide linear gains in air interface capacity with increasing number of femto-APs in a hybrid deployment. While the macro network provides coverage, small cells (pico and femto) are better suited for capacity infill and indoor coverage. Small cells require no tower infrastructure or low lease cost, therefore drastically cutting the operational and capital expenditures.

Despite the benefits HETNETS pose a new set of challenges for network planners including:

- Cross tier interference: A dense femtocell deployment poses significant interference to macro cells. While interference to data can be addressed via intelligent use of Fractional Frequency Reuse (FFR), interference to control signals requires new mechanisms.

- Mobility management: Handover across small cells at moderate to high speeds gobbles network resources, and can degrade user experience if not managed well.

- Self-Organizing Networks (SONs) are essential for consumer deployed nodes like femtocells, and are important for managing inter-tier deployment.

- Security management between nodes of different ownership (consumer, enterprise, operators) and Service continuity, QoS management and delivery across multiple tiers are essential for high performance and high capacity heterogeneous networks.

Backhaul is needed to connect the small cells to the core network, internet and other services. Mobile operators consider this more challenging than macrocell backhaul because small cells are typically in hard-to-reach near street level rather than in the clear above rooftops and carrier grade connectivity must be provided at much lower cost per bit In one survey, 55% operators listed backhaul as one of their biggest challenge for small cell rollout.Many different wireless and wired technologies have been proposed as solutions, and it is agreed that a 'toolbox' of these will be needed to address a range of deployment scenarios. An industry consensus view of how the different solution characteristics match with requirements is published by the Small Cell Forum.Adding Wi-Fi to a small-cell enclosure does not increase its price significantly – and equipment typically accounts for a small part of the capex. The higher backhaul requirements may lead to an increase in equipment cost, but that is unlikely if operators use fiber or high-capacity wireless backhaul.

Injecting Wi-Fi to cellular small cells not only gives the small cells a capacity boost, it substantially lowers the TCO for the combined cellular and Wi-Fi deployment, because the marginal cost is low.The same is true in the reverse: adding cellular radios to operator-owned Wi-Fi deployments likewise lowers TCO, because the marginal cost of adding LTE to a Wi-Fi hotspot with sufficient backhaul capacity is low.Furthermore, combining Wi-Fi with 3G small cells makes it more attractive for operators to have initial 3G and Wi-Fi small-cell deployments, in networks without LTE or without any need yet for additional 4G capacity, but with severe 3G congestion.

Equipment installation and operating costs remain the same when adding Wi-Fi to a cellular small cell, because the siting, installation and operating requirements are the same. As Wi-Fi and cellular become integrated in the same small-cell enclosure because of the TCO savings, a decision between Wi-Fi and LTE is no longer needed, because both get deployed throughout the small-cell footprint. Deploying a small cell with an integrated Wi-Fi solution is part of an overall strategy that operators will find maximizes the investment return from small cells. These additions leverage the investment of small cell deployments and require an incremental CapEx to add Wi-Fi into the solution mix.

China Mobile and AT&T are among those incorporating Wi-Fi into their heterogeneous network architectures and expect more operators to follow suit. Hotspot 2.0, or Passpoint, is also gaining traction – Time Warner Cable recently announced a national deployment of the technology for users of its Wi-Fi network, which means that users only have to log into the Passpoint network once and are automatically authenticated whenever they are in range thereafter. For outdoor cells, the main issue now appears to be the ability to acquire the sites where the mini base stations should be deployed – such as lamp posts and bus shelters – rather than with any technical issue. Some Pundits recommend the "crowd-sourcing" model, while operators say they are open to working with aggregators and the cities themselves.

When planning for a capacity expansion, the use of operator-owned carrier Wi-Fi networks will reduce the cellular capacity requirements. When planning for a small-cell sublayer, operators have to carefully select locations both for Wi-Fi and for cellular small cells where the subscriber density is highest. An even or peaked distribution of subscribers within the macro footprint, the choice between indoor and outdoor coverage, and the area topology have a great impact on the capacity gain that Wi-Fi and cellular small cells contribute. There is no doubt that Heterogeneous networks will enable the cost effective deployment of high performance networks in order to bring wireless broadband to every corner of the globe.

In future one expects client clustering and P2P communication to transmit/receive information over multiple paths between the base station and client. This creates the potential for improvement in throughput, capacity and reliability without increased infrastructure cost.The adoption of HetNet RAN architecture has a big impact on network management. In particular in order to manage the interworking and interoperability between macro e micro layers, it's fundamental to adopt a solution called SON (Self Organizing Network).The main aim of the SON suite is to perform the optimization tasks, routines and activities, usually manually carried-out by engineers, in an automated, continuous, autonomous and closed-loop way, applying pre-defined optimization policies and rules.

SON can be thought of as a toolbox of solutions. Operators can choose which ones to adopt and when, depending on their needs, their strategies, and the maturity of the solutions. SON standardization efforts started with 3GPP Release 8 but are still ongoing, so there is a varying level of maturity among tools, in terms of both specifications and the commercial availability of products. The focus of SON standardization has gradually moved from the macro-cell layer to the small-cell layer, as the small-cell market expands and ecosystem players encounter the challenges that small cells introduce in mobile networks.

The transition to HetNets has already started. In fact, it could be argued that it started even before the term was coined – mobile networks' complexity started to grow once 3G (and, with it, multi-RAT environments) was launched – but we are still seeing only the initial impact HetNets will have on how mobile operators run their networks. Even so, operators and the vendors working with them are on a steep learning curve, and we have clear indications of the key challenges they face when deploying HetNets.

---♠---

2.5: Carrier Wifi : Solution for the African broadband crisis ?

Despite the arrival for so much submarine cable bandwidth it seems the hapless African broadband consumer will not experience true high speed broadband for atleast a decade. So what is the current situation with the Operators in South Africa ? Congested networks , refarmed 1800 spectrum for LTE which is ruining the HSP + experience , smartphone flood , insufficient Spectrum , disgruntled customers , declining rather than value based pricing etc etc. Not to mention the fact that some Operators become professional spectrum squatters without the financial means to build a national broadband network. As for the rest of Africa : the same lamentable situation.

The Regulators are stubborn since they refuse or delay the allocation of Spectrum in the 2.6 and 800 mhz for LTE. The Operators are asinine since refuse or delay in look at non licensed Spectrum to rollout a RAN. Infact the Regulators and Operators spend more time in criticising rather than collaborating with each other. The Digital Dividend is a pipe dream for the majority of the African population. Most Africans don't have access to electricity or clean water what in the world are they going to do with Broadband ?? Nobody gives a hoot !

The profitability of mobile Internet services is at risk, unless mobile operators and Regulators look to new, complementary technologies that can serve data traffic with a dramatically lower cost curve than the traditional, cellular-only RAN and backhaul solution. It takes many years and billions of dollars for mobile operators to acquire and build out new licensed spectrum. Operators cannot simply double their use of spectrum and double the number of cell sites in their network each year.If no changes are made demand could exceed capacity within two to three years and with today's architectures operators could rapidly run out of spectrum and money. Operators need to harness a solution that will further dramatically bring down their backhaul and RAN costs, while allowing them visibility and control of the new RAN, maintaining the same ease of use for their subscribers.

The Wi-Fi offloading has became lately the most hotly debated business opportunity that provides solutions for MNOs for a lot of challenges like spectrum licensing, running costs, coverage gaps ,deployment delays, and congestion. In addition, Wi-Fi can give new business opportunities to MNOs to access new types of users like laptop, tablet, and home users.Offloading to Wi-Fi is the fastest way to achieve this dramatic reduction in backhaul costs (compared to fixed broadband) and RAN costs, without the need to purchase additional spectrum. Wi-Fi offers an immediate solution to operators worldwide looking to increase capacity and coverage. Its availability is widespread and growing, with over one million commercial hotspots today. Unlike cellular radio technologies, Wi-Fi uses the same frequency bands worldwide and users have had few problems sharing this unlicensed spectrum.

Over the years Wifi has undergone a resurrection of sorts with new standards that integrate it more tightly into the Mobile Core. The revolution of smartphone industry, which rose after the Apple iPhone®, had a vast spread of Wi-Fi technology along with smartphones. The 3rd Generation Partnership Project (3GPP), a collaboration between groups of telecommunications associations aims to make a globally applicable third-generation (3G) mobile phone system specification, became aware of the increasing role of Wi-Fi and started putting standards for Wi-Fi and Mobile Networks interoperability. Some mobile operators, Wireless Internet Service Providers (WISPs), and vendors saw the potential in Wi-Fi as generic and low cost wireless technology to provide their data and Internet services to millions of users through their already equipped Wi-Fi in smartphones, tablets, laptops, and PDAs.

And yes : Wifi is now CARRIER GRADE meaning ready for use by Telco operators !!!

• The Wi-Fi Alliance Passpoint certification program and the Hotspot 2.0 specifications provide seamless Wi-Fi access in public hotspots and when roaming. With SIM-based authentication, mobile devices can automatically connect to any hotspot operated by a mobile operator or any of its partners, as they do with cellular data roaming.

• The Wireless Broadband NGH initiative provides a roaming framework that facilitates roaming agreements among mobile and Wi-Fi operators, and establishes roaming best practices for Wi-Fi.

• ANDSF facilitates discovery and selection of non-3GPP networks in mobile devices. With ANDSF, operators can use PCRF-defined policies and real-time traffic management across mobile and Wi-Fi networks.

- 802.11n : 802.11n is an amendment which improves upon the previous 802.11 standards by adding multiple-input multiple-output antennas (MIMO). 802.11n operates on both the 2.4 GHz and the lesser used 5 GHz bands. It operates at a maximum net data rate from 54 Mbits/s to 600 Mbits/s.

A new IEEE standard is in the wings ready to make its debut on the wireless market stage: 802.11ac. In essence, 802.11ac is a supercharged version of 802.11n (the current WiFi standard that your smartphone and laptop probably use), offering link speeds ranging from 433 megabits-per-second (Mbps), all the way through to multiple gigabits per second. To achieve speeds that are dozens of times faster than 802.11n, 802.11ac works exclusively in the 5GHz band, uses a huge wad of bandwidth (80 or 160MHz), operates in up to eight spatial streams (MIMO), and a utilizes very fancy technology called beamforming.

Hotspot 2.0 was created by the Wi-Fi Alliance in 2012, a technology intended to render Wi-Fi technology similar to cellular technology with a suite of protocols to allow easy selection and secure authentication. It allows the mobile devices automatically select the Wi-Fi network based on its SSID. It also allows reacting some useful information such as Network and venue type, list of roaming partners, and types of authentication available.

The 3GPP provided further enhancement with release 10; a completely seamless Wi-Fi offloading, where the mobile device can have multiple connections to each technology managed by the 3GPP core network. Some heavy traffic like video streaming and P2P downloads can be routed via Wi-Fi and the HTTP and VoIP traffic through the cellular Network.

Wi-Fi is expanding its role in carrying wireless broadband traffic in both private and public networks. As it becomes more deeply integrated into the mobile network infrastructure, Wi-Fi will provide the same seamless experience that subscribers are used to in cellular networks. Automatic SIMbased authentication, optimized network selection, secure connectivity, and policy-based service options will increase the value of Wi-Fi access even as operators move to higher-capacity, faster LTE networks.

WiFi data offload can alleviate mobile network congestion, reduce churn and lock in customers with bundled WiFi/3G services. It can reduce operating expenses through the use of lower cost infrastructure, and increase revenue through subscriber retention and increased market share.Wi-Fi roaming also delivers important benefits to operators. Analysts agree that operators will greatly increase the amount of data traffic they offload to Wi-Fi and femtocells. They estimate that 30% to 60% of total data traffic will be offloaded from 3G and 4G mobile networks by 2015. At this rate, by 2015 commercial Wi-Fi networks will carry as much mobile data traffic as was carried by all mobile networks in 2010.

Leading Mobile Operators are using carrier grade Wireless Mesh as part of the Data Offload and incremental revenue strategy. In 2010 Mobily Saud Arabia (who launched the first TD LTE commercial network in the world) noticed a massive increase in data usage, fuelled by offering unlimited data subscriptions. Against this backdrop of increasing data usage, Mobily saw Wi-Fi as an efficient way to reduce the cellular capex investment in broadband infrastructure needed to match this spike in data usage. Today Mobily has around 350-400 public hotspots with each Hotspot comprising of multiple Wi-Fi Access Points covering multiple business verticals including cafes, hotels, hospitals, outdoor, and some challenging venues such as stadiums and Holy Pilgrimage areas.

Mobily's business model is predicated on a hotspot portal based Wi-Fi virtual network with multiple service monetization models for both Mobily and non Mobily subscribers PLUS a cellular-to-Wi-Fi offloading virtual network, offering seamless and secure user experience with the use of EAP-SIM protocol and WISPr clients. Mobily intends to "offload" at least 20% of current mobile broadband traffic onto Wi-Fi networks and is designing the Wi-Fi network to meet this key performance objective . The Hotspot 2.0 standard will open the door for inbound roamers to connect seamlessly and securely to Mobily's Wi-Fi network while their usage is being charged back to their home operator.

The devices using the most mobile data – smartphones, laptops and tablets – all support Wi-Fi. Currently, laptops and tablets ubiquitously support Wi-Fi and support of Wi-Fi by smartphones will increase from 50% to 95% by 2014.With smartphones and Broadband 'dongles' consuming considerable volumes of data very easily bill shock refunds have plagued Mobile Operators in advanced countries. As such, communications such as that sent by AT&T, particularly to iPhone users, advising them turn off Data roaming and suggesting the use of Wifi alternative have been quite common.

A seamless Wi-Fi offering will provide an attractive up-sell opportunity to existing mobile data bundles; either as a flat-fee, mobile roaming charge (MRC), or perhaps a free feature that can then be charged through usage. A good data offload solution can potentially play a wider role in extending service support to other technologies such as WIMAX or LTE, e.g. providing support for interworking of voice and messages services between 2G/3G and LTE. Essentially, capex can replace opex in the short term if it is justified by a longer-term return on investment. Therefore, 'LTE readiness' becomes crucial from a corporate as well as an operational perspective.

TIM Brasil, a pioneer of Wifi SIM authentication, started the process of rolling out a large-scale carrier-class WiFi network to offload 3G and to reach previously underserved communities in the cities of Rio de Janeiro and Sao Paolo. TIM Brazil's solution offers automatic EAP-SIM authentication to smartphones and

SIM-enabled tablets, working joined with mobile network in order to manage the traffic bundle of the customer.WiFi was also installed in Airports and Stadiums; in April 2014 during the match between Botafogo and Union Espanhola, with 44.000 people watching in the Maracanà, around 28GB of traffic were managed by TIM WiFi network.In all the World Cup's matches the WiFi offload was always between 20% and 45%, helping mobile network to provide good performance to the supporters.

So what about reliable high speed Broadband for the masses ? Don't be surprised if it is not LTE running on 800mhz because the allocation of Digital Dividend has been delayed in Africa : It might be Wifi backboned on terrestrial fibre (esp in the rural areas) to give the the african subscribers what they deserve after years of patchy low speed connectivity languish !!

---------------------------------♠---------------------------------

Chapter 3 : Planning a Digitised ICT Infrastructure

3.1 : Digital Infrastructure : LET'S GO DUTCH

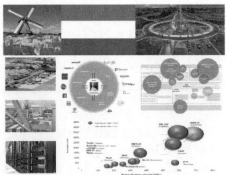

One of the few countries in the world that punches way outside its weight class is Holland aka the Netherlands.Comparatively, the area occupied by the

Netherlands (33,883 sq km) is just twice the size of Gauteng (the economic hub of South Africa).The 16 million Dutch people pump out a whopping GDP that is almost 3 times South Africa with its 53 million people. Unlike resource rich South Africa , Holland has nothing except the diligence and intelligence of its people. From an early age there's a feeling instilled in the Dutch that they can do anything. You can cycle anywhere in the land without fear of being laughed off the road or run down by speeding maniacs.

The Dutch transport system is brilliant : trains run all night long between Amsterdam and the Hague. Not to mention the delish cheese , pictoresque canals, a national park with an amazing museum hidden inside and the breath taking painting masterpieces at Rijksmuseum.... well you'll just have to go visit to see the rest !

Ofcourse Holland is not all cheese and tulips and its great people who gave military funerals to all the victims of MH17 flight tragedy over Ukraine.In Europe, the Dutch are among the frontrunners in the area of Digital Infrastructure (Internet connectivity, colocation housing and hosting).In many ways this infrastructure fulfills a gateway function, similar to that of Schiphol Airport and the Rotterdam Harbor. Amsterdam based AMS-IX remains the largest Internet exchange point in the world and the colocation housing market, centered around Amsterdam, produces strong growth rates. The Dutch also rank among Europe's elite in hosting. The services that encompass Digital Infrastucture on core Internet on one hand and colocation, hosting and IaaS on the other include :

- Internet Exchange: Parties that facilitate networks to interconnect with each other to exchange Internet traffic mutually (peering). This is typically done without charging for the traffic
- Transit Provider: Parties that provide network traffic in the 'core' Internet and connect smaller Internet service providers (ISPs) to the larger Internet
- Colocation: Delivering facilities (floor space, power, cooling, network connectivity) to enterprises and service providers for

housing servers, storage and other computer equipment as an alternative for an in-company data centre
• Dedicated hosting: Delivering computing power and storage via equipment dedicated to a specific client but managed by the hosting provider
• Shared hosting: Delivering computing power and storage by sharing the resources of physical equipment among multiple customers
• IaaS: Infrastructure-as-a-Service, delivering computing resources (e.g. servers, storage) according to a model that meets the essential characteristics of Cloud computing: on-demand self-service by the customer, measured service (pay-per-use), rapid elasticity (any quantity at any time), resource pooling (multi-tenant model) and broad network access (infrastructure is available over the network via standardised mechanisms

The Amsterdam Internet Exchange (AMS-IX) is the largest in terms of connected Autonomous System Numbers (ASN). The significance of an Internet Exchange is measured by the number of peering networks (Autonomous System Numbers) and and the Peak Internet traffic in Gigabit per second. AMS-IX is a mainport for Internet traffic more than Rotterdam and Schiphol are for containers and passengers respectively. London, Frankfurt, Paris and Amsterdam form the leading group of colocation data centres hot spots in Europe. Measured in colocation supply m2 per € bn GDP, Amsterdam exceeds all other cities.As a result Netherlands is hosting the top of the world's technology and Internet companies as gateway to Europe and the Internet such as : Facebook , Twitter , Netflix , Akamai , Amazon , Google and on and on. For this reason world's largest service providers and e-commerce companies have chosen Amsterdam as their #1 or #2 Internet Exchange position in the EU.

Large investments in data centres within the Netherlands by corporate multinationals like Google and IBM generate additional employment. Direct employment in the Digital Infrastructure sector adds up to 7,600 FTE, of which 90% in the hosting sector and 10% in capital intensive housing. Operational expenditures and investments in the housing and hosting sectors drive indirect and induced effects to create additional jobs. Combined effects for the Digital Infrastructure add up to 19,000 jobs in 2013 with a projected growth of 8% a year.The real value of the Digital Infrastructure sector, however, lies in its significant impact on the much larger Internet economy and broader digital society.

A continues interaction between Digital Infrastructure, service innovation and online usage drives growth in the online ecosystem. Digital Infrastructure is part of a much larger online ecosystem generating at least ~ €39 bn in revenue in the Dutch economy. Including private investments, government spending and trade, the Internet economy in Netherlands adds an estimated €34 bn to the GDP which is approximately 5.3% of the total GDP and steadily growing. There is a strong correlation between the Digital Infrastructure and e-commerce which shows that the former is a key enabler because E-commerce application are hosted in data centres and e-commerce traffic flows over the Internet exchange(s).

The employment generated by e-commerce in Netherlands is estimated between 100,000 and 140,000. SaaS and PaaS are two of the Digital Infrastructure's closest relatives, generating 5,700 jobs in the Dutch economy. Google has invested €600 million on a data centre located in Delfzijl, the Netherland.The estimated additional employment that the data centre will provide is 150 FTE from operations and a 1000 FTE at the peak of construction. The presence of most major global data centre providers in the Netherlands is prove of the country's attractiveness in terms of Internet Connectivity, availability of required electricity capacity , economic and political stability and highly-educated and multilingual workforce.

A large and complex ecosystem of companies and other entities compete, collaborate and cooperate to construct and maintain the interconnected network of networks that is the internet. The ecosystem works, and anyone can download a web page or video, or activate a mobile app, because of common standards and a shared understanding among participants of the benefits of a vibrant and growing economic system. Countries need energetic digital service sectors. They are drivers of social and economic development, job creators, talent magnets and the exports of the future. Robust digital service sectors depend on a complex ecosystem that includes adequate infrastructure and an investment-friendly business environment.

The availability of mobile spectrum is one of the biggest, and most complex, infrastructure constraints especially in Africa. Governments must release additional spectrum – licensed and unlicensed – for private mobile use, as well as take steps to encourage spectral efficiency. New approaches to encourage harmonization and alternative deployment models are required.

BCG Reports that Emerging market consumers are embracing the mobile web as much more than a purveyor of convenience; they are using it to improve their well-being, intellect and earning ability. However the lack of broadband penetration in emerging countries – especially fixed, but also mobile – is a serious impediment to accruing the benefits of a first class Digital Infrastructure such as the one in Netherlands. This ought to represent an opportunity – many emerging markets are free to adopt new technologies, such as LTE and fibre, without the burden of managing legacy infrastructures. Progress has often been slow, however.

India, for example, has struggled to develop digital infrastructure. Fixed broadband reaches less than 10% of households, and while mobile penetration has hovered around 75%, it is dominated by 2G networks; 3G and 4G penetration is less than 5%. There is also a strong urban-rural divide, with mobile penetration in urban areas topping 160% while in rural areas it does not reach 40%.Indian mobile operators struggle with fierce competition, low consumer spending power and poor spectrum management. In Africa the mirror situation is even more pathetic !!

In the digital era, connectivity counts. It is impossible to imagine the country, sector, industry or area of endeavour that cannot benefit from digital services. The services enabled by digital technology are economic growth drivers, job creators, talent magnets and big sources of exports. The economics of many emerging economies make infrastructure (as well as other) investment tough. At the same time, a growing number of governments, companies and organizations recognize the benefits of expanding internet access as widely as possible. They also see that gaining access can have an outsized impact for people who live in particularly poor and remote areas.

Bridging this divide may require non-traditional and innovative approaches. Internet.org is a partnership started by a group of major technology companies (the founding partners include Ericsson, Facebook, Mediatek, Nokia, Opera Software, Qualcomm and Samsung) with the goal of working with governments and NGOs to bring basic internet services to people who do not have them. The underlying philosophy is that demonstrating the internet's value for free will cause users to want to pay for more or better services down the road (which is not too far removed from how internet use evolved in the rest of the world).

The policy-makers of the future (especially Developing countries) must be able to tackle the challenges posed by the digital economy. They need to consider the impact of policies on the entire value chain, including telecommunications, digital services and media, and ensure that any regulations that are deemed necessary are applied with a light touch and restraint. Perhaps most importantly, policy-makers need to take into account how quickly technologies and the innovations they enable are evolving.

Perhaps , start learning from the amazing Dutch how they did it in the Netherlands

---

3.2 : Formulating national ICT Policies : A blueprint for success

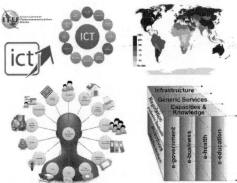

Many Governments in the emerging markets are highlighting the pivotal role of ICT in their socio-economic development. The main objective of national ICT policy is to balance the benefits and the risks of expanded ICT use in a way that is consistent with national development goals. The services sector has become increasingly ICT-intensive, and the knowledge sector is largely dependent on ICTs. ICTs enhance productivity across all sectors, including government.

In the past, there have been too many examples of the ICT arriving first, with the solution looking for a problem.Technology is an enabler and ICTs can become an enabler but it is imperative that this is guided by an information society strategy.National ICT policies typically concern themselves with readying the country, its economy and society for the information society. This usually cuts across a range of sectors, including education and health, finance, small and medium business and government (e.g., developing e-government capacity and services). The move towards an Information Society is distinguished by the following characteristics:

• Growing dependence on ICTs: As ICTs become more pervasive in business and personal contexts, people become more dependent on them for their livelihoods and for fulfilling their social and recreational needs. Being unable to access or use ICTs can become a serious deprivation;

• Growing ICT sectors: The provision of ICTs and related services forms a sizeable sector of many economies. Increasingly, developing countries are introducing high-level ICT strategies that aim to develop this sector of their own economies as well as using ICT as a tool in other sectors. One study for a mobile network operator has suggested that a 10 per cent increase in mobile penetration in a country can grow the gross domestic product by 0.6 per cent.

• More use of ICTs: Economic development and growth entail a shift in the proportions of national output, away from the primary sector of agriculture, through the secondary and tertiary sectors of industry and services, towards the new information economy.

ICT policies can be integrated into sectoral as well as broad national policies; for example countries may commit to introducing ICTs into schools in order to expand educational opportunities and increase the supply of ICT-literate graduates; they may extend internet access to rural clinics to improve the delivery of health services. As the use of the internet expands within countries a host of specific issues emerge: privacy and security, intellectual property rights , access to government information are examples.

The World Summit on the Information Society (WSIS) objectives and the Millennium Development Goals (MDGs), include a global partnership for development whose target is to provide citizens with all the benefits of new technologies, especially information and communications, in cooperation with the private sector. The development-strategy-led approach now points to the inclusion of ICT goals in Poverty Reduction Strategy Papers, in order to ensure the availability of ICT as and when needed for poverty reduction. There is some debate around whether and how ICT deployment assists in reaching the MDGs, but the following points seem clear:

• ICTs can help in implementing many initiatives that contribute directly to reaching development goals even when they do not necessarily contribute directly themselves;

• ICTs have impact that depend on the technical, economic, administrative and social environment, so general assessments of their contributions without considering the local context are difficult;

• ICTs are increasingly understood to be complementary to other development imperatives and not to be traded off against them.

The future regulation of the ICT sector depends upon reliable studies (using best practice research and analysis toolsets) into the ICT needs of the population and the maintenance of a database of relevant statistics. Such research/analysis must achieve (but is not limited) to the following key objectives:

• Evaluate the usage of ICT by various segments of the
population: Households , Enterprises and Government institutions

- Create a reliable database of statistics on ICT usage so as to measure the impact of Government policy measures , and guide future development of ICT and Telecommunications via regulatory interventions
- Supplement the current statistics provided by the telcos on fixed and mobile Internet usage to assess the usage of ICT services by private and public sector.
- Examine various developments in the rollout of broadband infrastructure (wired and wireless) to model the impact on ICT diffusion in the country.

The research study must encompass a number of global best practice methodologies to constitute the underlying data and analysis necessary for the development of ICT usage database such as :

- Background study and database: This is a detailed desk study that summarizes the geographic, demographic, socio-economic and cultural composition of the country. Ideally, the database and analysis should be broken down to the smallest local administrative level for which it is feasible to collect data. This is often at the district level, but in populous countries, or ones where data is freely available, data to sub-district level is desirable. Household income and expenditure data is especially useful in evaluating the ICT status.

- Telecoms and ICT sector review: This encompasses an inventory of existing infrastructure and services around the country, but also includes a review of the policy and regulatory environment for ICT, and possibly even the investment and business environment. Usually, the best approach is to interview the ICT industry players directly who will provide data on current network services and reach, as well as future plans, views on market trends, and their opinions on universal access, rural communications and progression towards universal service .

- Coverage and GIS maps: The information gathered from reviews and studies can be represented with coverage and GIS maps. However, because the ICT market is evolving rapidly, data can quickly be out of date. The focus of the ICT sector review should be to enable a policy formulation based on an understanding of the current situation and near future developments; it does not require absolute accuracy. It is nevertheless, helpful to set up a process and structure that allows for regular reviews (e.g., annually) of the ICT sector and related data.

- International review: Policy makers benefit from researching and discussing current best practice and trends for ICTs , especially of countries that have comparable characteristics and challenges.

- Demand studies: These are particularly valuable as they gather information from the intended beneficiaries of the ICT policy in regards to their actual needs. By investigating affordability, crucial information is gathered to model the subsidy requirements for various ICT objectives.

Ultimately any research study must satisfy the Government's need to create a legal framework that will direct investment into the development of ICT services in the country.In any ICT development initiative the following parameters must be factored into the final recommendations :

- Regulatory measures that create an environment more conducive to competitive network expansion or infrastructure sharing;
- Fiscal measures that will make communications service and hardware more affordable to low-income users;
- Enabling activities, such as promotion, advertisement and capacity building that highlight the opportunities available to

people, communities and organizations to take advantage of the services offered in the competitive market.

Manuel Castells, in his three-volume work on the information age , has suggested that a networked society is one in which "the entire planet is organised around telecommunicated networks of computers at the heart of information systems and communication processes." This dependence on the power of information reaches us all ". Furthermore he states , "the availability and use of information and communication technologies are a prerequisite for economic and social development in our world. They are the functional equivalent of electricity in the industrial era."

Castells goes so far as to conclude that ICTs can allow countries to "leapfrog stages of economic growth by being able to modernise their production systems and increase their competitiveness faster than in the past."

---------------------------------------♠---------------------------------------

3.3 : The Smart Connected life : Telcos are in pole position

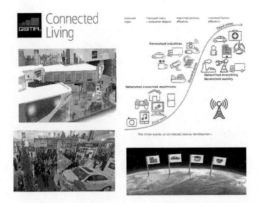

According to GSMA "Beyond connectivity, mobile operators will play a crucial role in working together with a range of industry partners in health, automotive, education, smart cities and a range of vertical industries to accelerate the launch of valuable connected services," Unfortunately without continued investment and growth in mobile networks (especially LTE) and the deployment of multiple connected devices, the socioeconomic benefits of the connected life will not be fulfilled.

mHealth programmes are currently one of the most cost-effective ways of providing remote living assistance to aging and chronically ill patients. mHealth programmes provide faster response times, integrated record access and considerable ease of use to patients. Remote consultation and support is expected to address the growing chronic disease management issue by reducing the need for hospitalisation. Proactive mobile based care for patients with sudden health incidents can reduce the number of primary and emergency visits by 10%. Mobile technology can also be used for home monitoring, thereby reducing the need for face-to-face consultations.

This year in Barcelona MWC , Telcos had the opportunity to generate value beyond basic connectivity through managed connectivity, stewardship services and platform innovation. The GSMA area was filled with interactive demonstrations of the connected life including the Aston Martin One-77, the bike of the future. It is fully connected and tuned into its own performance as well as the rider, including mobile health monitoring and electronics that track the bike's performance in relation to its environment.

There was also the Mantarobot showing virtual teaching through augmented reality and virtual presence and the Cooltra Connected Electric Scooter, the latest in smart city transportation, the GO! S3.4 from GOVECS which lets customers know when and where scooters are available and can be started with a phone via NFC. The widespread penetration of mobile networks offers a powerful platform to improve access to relevant content.

mEducation solutions already allow thousands of students in China, Bangladesh, South Korea and Indonesia to access course content through SMS and audio lessons. An mLearning student saves 86.7% of the cost spent by students taking the same training in a traditional classroom. Much of this is due to the elimination of the cost and inconvenience of travelling to attend courses. Inexpensive personal learning devices like the 35 USD tablet launched in India are further improving access to mEducation.

In developed countries mobile interventions could help cut healthcare costs by 400 billion USD in 2017, help retain 1.8 million students in the education system, save one in nine lives lost in road accidents, and reduce CO_2 emissions by 27 million tonnes annually. Similarly in developing markets, mobile interventions could help save over a million lives in Sub-Saharan Africa, provide education access to 180 million students, save 25 million tonnes of food and encourage over 20 million commuters to start using public transport. (GSMA Connected Living Program)

Mobile networks play a pivotal role in the development of the connected life providing a scalable, standardised global platform to support the growing demand for intelligent, secure connectivity.Examples of valuable connected services were amply demonstrated by leading edge Telcos at the GSMA Connected City showcase in Barcelona included :

In the Connected Home AT&T showed how people can use their smartphones and tablets to manage their energy, automate appliances and secure their homes through AT&T Digital Life. General Motors demonstrated how AT&T's 4G LTE network will transform the driving experience by enhancing safety, security, diagnostic and infotainment in the vehicles starting next year.

Deutsche Telekom, in conjunction with IBM, bought to life Smarter Cities for the Future using machine-2-machine technology to optimise urban services such as public transport, parking, energy, security and water management. Together with SAP, Deutsche Telekom also showcased Connected Port Solutions designed to optimise both road and sea traffic control as well as logistics and terminal operations in order to make port processes more efficient allowing larger quantities of goods to be trans-shipped in the port area.

Korea Telecom featured technologies that make our lives better including edutainment robots, automatic content recognition, smart home phones, a controlled motorcycle, eco food bins and cloud CCTV.There were also Smart Apps in your hand showing how we can live smarter with intelligent and unique applications including mobile K-pop music, integrated mobile payment and self-created M-learning solution.

Vodafone were showcasing their Smart Home, Smart City and Smart Mobility solutions.Smart Home illustrated how M2M technology can provide premium security services, enable remote health monitoring and even open and close doors remotely.Smart City demonstrated how Vodafone's Energy Data Management (EDM) solution, solar energy production monitoring, remotely controlled street lighting and digital signage are enabling the smart city.

Vodafone's Connected Cabinet solution demonstrated retail display cabinets that report on location, operational status and stock levels in real-time. Smart Mobility showed how M2M is transforming the automotive and transportation industries, be it through real-time information systems for public transport, enhanced drivers' experience with telematics services or through usage based insurance services with Vodafone Vehicle Connect.

---------------------------------------♠---------------------------------------

3.4 : National Broadband Policies : A Call to Action

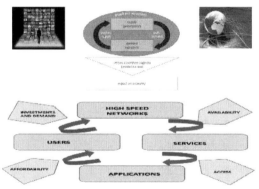

It is the goal of every African nation to adopt regulatory and statutory measures to promote affordable and widespread access to broadband services and to support the development of ICT skills and to adopt national e-strategies. The evidence is fairly conclusive that broadband has a positive contribution to GDP and that this contribution is magnified as penetration increases.

In recent years, several countries in the developed and developing worlds have formulated national broadband plans; these plans outline both coverage and service targets, as well as policies, with the purpose of achieving near or complete universal broadband service. National broadband plans touch upon four broad policy areas:

Assignement of Spectrum assets for universal broadband service coverage : It is also widely recognized that wireless broadband will be the primary technology used to provide ample broadband coverage in developing countries. As such Regulators in Africa must accelerate the allocation for high demand spectrum (2.6 and 800mhz) to set up LTE 4 G networks.

Investment in promotion of adoption programmes: is oriented at addressing demand gaps with particular focus on: universal service policies; the stimulation of the adoption of broadband through digital literacy; economic subsidies; deployment of public access centres; and the development of eGovernment applications in order to promote adoption of broadband.

Adoption of a competition policy: an endorsement of either facilities-based competition between vertically-integrated players such as the telecommunications operators or service-based competition (through unbundling of the telecommunication network of the incumbent operator and the sharing of incumbent facilities).

Removal of any potential supply obstacles: relates to the belief that competition among service suppliers is the right model to stimulate broadband supply, national broadband plans focus on how to lower economic barriers to entry. Relevant policy initiatives could include infrastructure sharing policies, which can range from stipulating rules for duct, mast, and tower sharing to lowering pole attachment costs (in aerial networks) to joint trenching rules.

With the rise of broadband-enabled services and applications and the increasing migration of many aspects of modern life online, a lack of broadband connectivity can increasingly have a negative impact on social and economic development by excluding those who lack broadband access or do not understand the relevance of broadband-enabled services.

A good plan should aim to promote efficiency and equity, facilitate demand, and help to support the social and economic goals of the country. The most successful plans will start with a clear vision of what broadband development should be and contain well-articulated goals that can be used to develop specific strategies to achieve success.

Any national broadband policy needs to fulfill the following imperatives :

• Orienting private investments to ensure wide regional coverage of broadband services;
• Making complementary public investments in basic transport infrastructure to promote competition in non-replicable network segments;
• Promoting service affordability and appropriate service quality benchmarks; and
• Stimulating broadband demand through complementary investments in digital literacy, content and applications, research and development, and public access centres.

It is important that the Regulator or Ministry Of Telecommunications (whoever has the mandate to compile a

National Broadband Plan) understand the challenges that result from well-documented market failures in the provision of broadband services. Among them are:

• Regional disparities in broadband penetration, which reduce development opportunities for the poorest regions
• Limited capillary presence of the backbone infrastructure for data transport, which results in high prices and low service quality outside the main urban centres;
• Limited connectivity among public schools, libraries, and government offices;
• Inadequate skills and low penetration of terminal equipment among disadvantaged households, which reduces demand incentives; and
• Limited development of local content and appropriate electronic services, which also reduces broadband demand.

The government of the Republic of Korea was one of the early broadband leaders. It has developed six plans since the mid-1980s that have helped to shape broadband policy in the country. The Korea example shows that policy approaches can effectively move beyond network rollout and include research, manufacturing promotion, user awareness, and digital literacy. As such in any broadband policy initiative the Govt needs to provide concrete actions rather than political bombast and take responsibility :

• Government should focus on maximizing competition, including removing entry barriers and improving the incentives and climate for private investment.
• Government should provide for specific, limited, and well justified public funding interventions only in exceptional circumstances (for example, where governments are trying to promote growth of underdeveloped markets).
• Government funding or policy should not compete with or displace private sector investment.

- Subsidized networks should be open access (that is, they should offer capacity or access to all market participants in a nondiscriminatory way)
- Government may need to regulate dominant providers to avoid market concentration or other adverse impacts on overall market competition.
- Government should eliminate barriers to content creation and refrain from blocking access to content, including social networking sites, or restricting local content creation.

In order to realize broadband's full potential for economic growth, an educated workforce trained in the use of ICTs is necessary. Additionally, there is a self-reinforcing effect between education and broadband, since broadband can help to improve fundamental educational outcomes, including learning how to use broadband better.

A review of 17 impact studies and surveys carried out at the national, European, and international levels by the European Commission found that the services and applications available over broadband networks improve basic educational performance . These studies found that broadband and ICTs positively affect learning outcomes in math, science, and language skills. In addition to facilitating basic skills, broadband improves the opportunities for individuals with ICT training, and such individuals generally have a better chance of finding employment as well as higher earning potential.

---------------------------------------♠---------------------------------------

3.5 : Smart Meters : Only smart Telcos can play this game

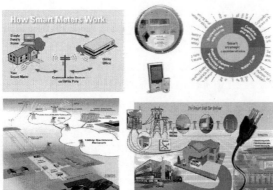

Next generation utility meters – or 'Smart Meters' – are a good example of the transformative potential of M2M technology. They will empower consumers by providing them with feedback on their energy usage, helping them to monitor, manage and – should they wish – reduce their energy consumption. Smart meters will also help reduce or end estimated readings, and make it easier for consumers to change tariffs and switch between suppliers, increasing market competition.

For the neophytes a smart meter is usually an electronic device that records consumption of electric energy in intervals of an hour or less and communicates that information at least daily back to the utility for monitoring and billing purposes.Smart meters enable two-way communication between the meter and the central system. Unlike home energy monitors, smart meters can gather data for remote reporting. Such an advanced metering infrastructure (AMI) differs from traditional automatic meter reading (AMR) in that it enables two-way communications with the meter. Smart meters are invariably part of an integrated program that pays for itself through reduced theft of electricity, energy savings,and operational efficiencies.

Equipping households and business with smart meters and collecting data on energy use enables better balancing of the network and a live view of grid activity, including technical failure. Granular, aggregated and analytical data generates information for better network planning and optimization to reduce losses.Data from the meters can be analyzed to detect unexpected consumption patterns, which can also indicate potential theft.Tamper controls can be built in to smart meters to alert utility companies to power theft.

The world's largest smart meter deployment was undertaken by Enel SpA, the dominant utility in Italy with more than 30 million customers. Between 2000 and 2005 Enel deployed smart meters to its entire customer base. These meters are fully electronic and smart, with integrated bi-directional communications, advanced power measurement and management capabilities, an integrated, software-controllable disconnect switch, and an all solid-state design. They communicate over low voltage power line using standards-based power line technology. Consumers are able to access their energy consumption and billing data online, and new methods of payment will be possible.

The overall revenue opportunity in smart electric metering is substantial, amounting to nearly $57 billion over the coming years, Navigant Research forecasts annual revenue will grow only fractionally, from $5.2 billion in 2012 to nearly $5.3 billion in 2022, with a compound annual growth rate (CAGR) of 0.1%. The Western European and Asia Pacific markets represent healthy growth potential, both in installation of new smart electric meters and upgrades to existing smart meter technologies. Driven primarily by China, the forecast penetration rate in Asia Pacific will reach nearly 70% by 2022. Growth in Europe is due largely to the EU smart metering policy which calls for 80% of households to have smart gas and electric meters installed.

The World Bank estimates that Africa needs to build an additional 7,000 megawatts of new generation capacity each year to meet suppressed demand, keep pace with projected economic growth and provide additional capacity to support the rollout of electrification.A push to introduce smart meters could help to relieve Africa's stressed power supply and bridge this investment gap in the region's electrical infrastructure. Put simply, installing a prepayment smart meter guarantees that cash can be collected from a customer. This could provide the necessary missing link to secure revenues, thereby attracting global investors in new generation capacity and allowing investment in other critical energy infrastructure.

Modern energy systems built on top of smart meters offer enormous opportunities to nurture smart grid (a smart grid is a modernized electrical grid that uses communications technology to gather consumer data in an automated fashion to improve the efficiency, reliability, economics, and sustainability of the production and distribution of electricity) , smart home and e-health solutions, all currently in their infancy.Cellular is a technology that has proved that can scale economically future-proof smart energy systems and increased data traffic. It's secure, open and interoperable, and cellular's clear roadmap and continued development will support the Smart Metering deployments as the technology continues to evolve to focus on high-speed capabilities and rich data services.

A D Little point out rightly that in a saturated market, smart Telcos are now starting to move from mass-market products to more sophisticated and vertically integrated solutions as a means of addressing declining revenues in the residential telecoms market. However, their existing core competencies seem almost predestined for expansion into certain areas of Smart Grid allowing them to gain a foothold in the electricity market. The existing automated meter reading business has not tempted telcos, since the low level of data traffic on SIM-enabled meters makes average revenue per user (ARPU) unattractive.

To address above, the fundamental business model for automated meter reading needs to be broadened to include billing and management services for the energy market and the end-customer. Moving from a connectivity-oriented business, which represents only 10–15% of the value generated in this field, to a service provision model would allow operators to extend their share in the value chain up to 60–70%. As telecoms operators are currently developing their skill sets and platforms, it seems it will be only a matter of time before they move en masse into this highly attractive area.

Deutsche Telekom is counting on getting an extra €1 billion in revenue annually by 2015 from networking services targeted specifically to the energy, automotive, health and media industries. Deutsche Telekom's plan is to sell a fixed-price service that includes installation and operation of a communications box that transmits the smart-meter data to a secure data center every 15 minutes, computes it and then forwards it to the utilities. It also provides utility customers with the ability to view their power usage via a secure Internet site.

The UK's Smart Meter Implementation Programme is a major national infrastructure project that will involve the roll out of 53 million gas and electricity meters across the UK by 2020. Telefonica's proposed communications solution for the smart metre network is based on its existing cellular rollout, supported by am IPv6 based wireless mesh solution which will connect meters in areas without cellular coverage. The system is based on the 6LoWPAN initiative founded by an IETF working group dedicated to pushing IPv6 over Low power Wireless Personal Area Networks under the idea that "the Internet Protocol could and should be applied even to the smallest devices." The IPv6 mesh network can then be established over any available carrier, including but not limited to, 2G/3G cellular, wifi, power line IP, and Bluetooth.

Telcos face an interesting business opportunity and they can become one of the few players able to gain access to a share of the entire energy management market. Their ability to utilize their existing network infrastructure and enabling capabilities to manage critical applications much more effectively will be a strong point of differentiation compared to application providers without their own network infrastructure. In CRM and billing, there are many similarities between telcos and utilities, particularly when smart meters and flexible tariffs are taken into account.

Moreover, existing customer relations can be exploited as a sales channel, and bundling electricity metering with other telco products would be feasible (e.g. home or office automation, remote-management-over-mobile devices etc.). However, telcos would need to develop competencies in grid connection, utilities tariffs etc. The greatest challenge for telcos is to formulate a robust strategy and appropriate processes that allow them to act quickly and flexibly in response to the changing economic, technological and regulatory environment.

The bottom line : smart meters can create a "virtuous circle" linking energy companies, investors, customers, communities and regulators . With prepaid smart meters in place in homes and businesses, investors will be reassured that cash can be collected from customers to pay for investment in generation. Governments and regulators can fulfill their targets of controlling and optimizing energy consumption. Energy companies can improve customer service, run efficient network operations and dynamically manage supply and demand. And Telcos can utilise their vacated GPRS / Edge channels as their smartphone customers clog up the 3 G LTE bearers.

3.6 : Mobile Signature : Telcos , Bank , Government win for the people

The development of electronic signature in mobile devices is an essential issue for the advance and expansion of the mobile electronic commerce since it provides security and trust in the system. E-signatures provide security for the transactions with authenticity and integrity characteristics that make non-repudiation of the transactions possible. In many countries, such as Estonia, Germany, Singapore and Hong Kong, it has become a key element of e-government through its "utilization of wireless and mobile technology, services, applications and devices for improving benefits to citizens, businesses and government units .Driven by the growing surge for mobile interactions, mobile commerce and online digital purchasing, carriers worldwide are investing in mobile identity infrastructure as an economically efficient solution fraud detection/prevention and identity theft issues.

Many different technologies and infrastructures have been developed with the aim of implementing mobile signature processes. Some are based on the SIM card. Others work over the middleware of the mobile device and cryptographic providers. Finally, there are already some frameworks which are independent of specific mobile device technologies and make mobile signatures available to application providers.Mobile signature solutions can only work on compatible SIM cards, that match the WPKI specifications in terms of security and capacity, and contain a SIM Toolkit application capable of performing signatures. The PKI system associates the public key counterpart of the secret key held at the secure device with a set of attributes contained in a structure called digital certificate. The choice of the registration procedure details during the definition of the attributes included in this digital certificate can be used to produce different levels of identity assurance.

A solution must also be implemented on the operator side to manage signature requests. If the access control secret was entered correctly, the device is approved with access to secret data containing for example RSA private key.Security is guaranteed by cryptographic systems (e.g. SHA1) and on-board key generation. The service is only made available on EAL4+ certified SIM cards which provide a high level of security. Legal compliance is ensured by a country specific Electronic Signature Law that gives electronic signatures the same authentication level as wet signature as long as they rely on a "qualified certificate". Qualified certificates are defined by the ETSI Standards4 and a directive by the EU Commissions as certificates that are issued by an authorised Certificate Authority following face-to-face verification of both the user and government issued photographic identification.

According to GSMA , Turkey and Turkcell was the global first in launching a mobile signature service. The idea behind Turkcell Mobillmza was to offer a remote way to complete transactions equivalent to an "original" signature on a hard copy – making it possible to sign documents and authenticate oneself via a mobile phone,in a way that is legally approved, secure, easy and convenient.Their Mobile signature services are easy to use, since they don't require any software installation. The certificate is activated Over-The-Air once the user has subscribed to the service. Signature requests then automatically pop-up on the user's phone each time he requests access to secure services. Once the user has entered his PIN, the signature is sent to the service provider, who checks its validity and grants access to the service.

Turkcell Mobile Signature can be used in all transactions, except for the ones that require a ceremony and the witnessing of a third party such as marriage or buying a deed, that require a signature such as private affairs, public affairs, and banking affairs. For example, EFTs can be carried out over internet banking with mobile signatures. Turkcell has imported e-signature technology to the mobile realm and it contributes to the e-Turkey transformation by carrying all transactions that require signature to the virtual realm where you will not need new readers or smart cards.

It is often the case that service providers are reticent about adopting mobile signature solutions if there is not a large installed base of users, and users are not enthusiastic about services that are not backed by multiple service providers. This leads to a stand-off that can often threaten the commercial success of mobile signature services. Initially, Turkcell's project was supported by the five main Turkish banks, which together pushed for the government to adapt the electronic signature law. This collaboration helped drive adoption since banks offered customers pre-registration at their branches, and then sent the forms to Turkcell.The banks also promoted the use of mobile signature through marketing campaigns.

The initial business model for Turkcell Mobilmza was a pay-per-use model. The service was free to subscribe to, and users had to pay a fee each time they used the signature service. The idea was that the cost of the certificate would be covered after a certain number of transactions, and then profit would be generated by extra usage. But this model relied on consistent levels of usage from subscribers. However a significant proportion of non-active users made this model unsustainable. Therefore this business model was replaced by two complementary approaches:

- Monthly subscription: subscribers pay 5 Turkish Liras for an unlimited number of signatures
- Price per signature: service providers pay a small fee per transaction. Public enterprises and educational institutions are not required to pay this fee, because of their public service orientation. It is anticipated that service providers who actively promote mobile digital signature will also enjoy a waiver of this fee.

In Europe the Mobile ID program in Moldova is a government-led project that is being deployed in partnership with mobile network operators. It is designed to offer citizens the speed, privacy, convenience and transparency of digital access to numerous government services and information for citizens, including online applications and copies of official documents. Their selected UICC-based solution is compatible with all types of mobile telephones, whether feature phones or smart phones. The application allows citizens to confirm their identity and sign documents directly from their mobile phone, by entering a unique user-selectable PIN code. A Mobile ID solution is responsible for the entire life cycle management, from user registration to verification of mobile digital signatures, and connection to the Certificate Authority body and e-government portals.

Lattelecom offers the Mobile ID service platform to Latvian service providers and mobile operators. The Mobile ID users are able to securely sign in to online services and sign documents and transactions, simply by using their mobile phone. As mobile phones are typically always at hand, a legally binding digital signature can be done regardless of time or place.Lattelecom launched the service with nine service providers, including Latvijas Krājbanka bank, Riga city council, Lattelecom's and Latvijas Mobilais Telefons' (LMT) online customer service, along with local enterprises and universities. Lattelecom handles user and transaction validation, making it easier for other service providers to join in. Lattelecom offers the Mobile ID service to third-party service providers in a variety of industries, including other Latvian mobile operators.

The ability to leverage network assets, such as the Subscriber Database Management (SDM) system, and the potential for incremental revenue from third-parties such as credit bureaus, banks, and credit card companies, makes mobile identity a high priority service for carriers worldwide. Mind commerce thus expects that mobile identity infrastructure market will grow at a CAGR of nearly 17% over the next five years eventually accounting for nearly USD 12 Billion in revenues by the end of 2019.

Chapter 4 : Gearing up for M&A and Due Diligence

4.1 : Spectrum Valuation and Bid Strategy : How to do it right!

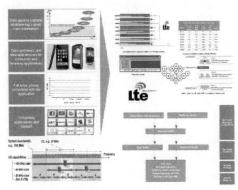

The release of additional spectrum is often used as a vehicle for introducing additional competition by the Regulators. From the perspective of spectrum regulators, careful spectrum management is required to ensure that sufficient spectrum is available to support not just the development of the commercial mobile market, but to support the continued operation of critical services such as Government, utility and Emergency Services that use radio spectrum on a daily basis. For the mobile operators spectrum is a critical resource, notably the ownership of lower band spectrum without which 4G will remain a myth in some emerging countries.

As we all know frequencies in the low band range 700MHz to 2.6GHz provide the optimal combination of propagation or coverage (the lower the frequency the better the coverage) and the ability to carry information or traffic (the higher the frequency the greater the data carrying capacity). Indeed some dense urban cities networks are now approaching the limit of network densification and additional spectrum and or new technologies may be the only route for alleviating network capacity constraints. Not to mention to stay in existence over the long haul !!

Unfortunately many mobile markets are no longer experiencing revenue growth as mobile broadband revenues simply offset declining voice revenues. To exacerbate the situation only smart

Telcos have figured out how to monetise LTE. Telcos will need to consider the requirements of these air interface choices – such as the levels of handset/terminal take-up, as well as base station, antenna and transmission upgrades – when formulating their spectrum acquisition plans. In emerging countries the demand for high speed data services and delays in availability of Digital Dividend Spectrum has caused severe congestion on their current 3 G networks.

Understanding the value of spectrum to a business is essential for developing a spectrum strategy and participating in a spectrum auction as rightly pointed out by the Coleago experts. The massive cash outlay for additional spectrum and the requirements to make a return on spectrum investments adds a layer of complexity to the evolving cost of data on HSPA and LTE networks. The substitutional nature of some spectrum bands requires a holistic approach to re-farmed 900 and 1800MHz spectrum strategy and valuation. The valuation process must consider stand-alone regional and / or block valuations and also packages of regions and / or blocks. When considering packages over stand-alone valuations the impact of scale must be included.

In respect of a spectrum auction an operator has to find an answer to three fundamental questions: How much spectrum do we need in different bands? The question relates to an assessment of spectrum need in the context of the growth in demand, notably mobile broadband. This needs to take account of the overall strategy, for example traffic offload through WiFi or Femto cells.

How much is each block worth, i.e. what is the most we should bid for it? This relates to valuing each spectrum block in order to set the bid limit for the auction. This is quite separate from auction strategy. Clearly, if there is no bid limit, the auction will be simple because a bidder would simply pay whatever it takes to win the spectrum. However, such an approach may not result in the creation of shareholder value and may draw criticism from shareholders and the financial press and capital markets.

How do we obtain the spectrum as cheaply as possible? In any auction, the bid limits should be set before the start of auction. The role of bid strategy is to ensure the spectrum is obtained for less than the bid limit and at the lowest possible price. Depending on the auction format there may be an opportunity to influence the outcome and avoid negative effects such as aggregation risk (e.g. be stranded with unwanted blocks). This is addressed by examining the auction rules and developing a bid strategy which will be tested through simulations and mock auctions.

Operators may also have to consider mitigating strategies for a "no spectrum case" but if they face network congestion a range of mitigating strategies such as traffic shaping and fixed line off-load must be examined and incorporated into the valuation process.Operators must consider how regulation on net neutrality might impact their ability to shape traffic profiles and whether there are any long run cost implications of offloading to other players fixed networks.

Spectrum is often awarded through an auction process and recent auction designs favour a second price rule which means that bidders cannot influence the price they pay for the spectrum only the price that others pay. A bidder's valuation for a spectrum block is the price at which he walks away from a take-it-or-leave offer. Where aggregation risk is present valuations should be defined over packages, not just individual blocks. A valuation is conditional on information known at the time.

Depending on the auction format there may be dominant bid strategies or ways to avoid negative outcomes in cases where there is aggregation risk. This can be explored at theoretical level, through simulations and mock auctions. In theory a bidder enters the auction well prepared and the auction itself is a mechanical exercise. However, as the auction unfolds there will invariably be some learning which needs to be processed at the end of each day in order to be prepared for the next day's bidding.

The next big Spectrum " land grab " will take place in Africa (the perennial laggard in the broadband era) even as the Regulators dither and delay in the implementation of the Digital Switchover / Dividend. We predict that LTE will really come of age in Africa in 2015 by which time the 700 : 800 mhz and 2-6 Ghz spectrum becomes available thru auctions or beauty contests. At a joint ITU (International Telecoms Union) and ATU (African Telcom Uniion) meeting the outcome saw Africa become the first region in the world to be in a position in 2015 to cohesively and harmoniously allocate bandwidth freed up by the transition to digital television— the so-called 'digital dividend'— to mobile services in both the 700MHz and 800MHz bands.

However if African Regulators dole out slivers of Spectrum to many " wannabe " Telcos in the false notion that this will decrease prices (as they did with Wimax) then expect the same mess : a host of under resourced " Pygmy " operators in each country setting up localised LTE networks with limited coverage and trying desperately to make a decent ROI. This scenario will do precious little to bridge the catastrophic Digital Divide in Africa. In a few years while the rest of the world will be on 5G African Telcos will still be boasting about their " 3 BTS me first 4G network " using their inadequate LTE spectrum allocations.

There is no doubt in my mind that African Regulators will split the LTE spectrum into slivers over many bands. But here is the good nesws : Carrier Aggregation (in LTE A benefits operators with multiple spectrum positions, those with small pieces, and particularly operators that are combining acquired networks. The initial focus is on higher-speed services, but expect more deployments of 5+5MHz carrier aggregation as emerging markets deploy LTE in 2014.By combining blocks of spectrum known as component carriers (CC) , carrier aggregation enables the use of fragmented spectrum and allows LTE-A to meet its IMT-Advanced headline data rate of 1 Gbps. In simple terms bonding Spectrum channels together to create larger channels enables faster wireless services, and reduce opex and capex costs from running multiple networks. Believe it or not LTE A + CA is the 4 G technology for emerging markets.

A Spectrum bid requires a well honed strategy that factors in technical , commercial and financial parameters to balance a subtle equation that underlines 4G data networks. This is followed by bid strategy that will be implemented systematically along a project time line by a " tiger team " drawn from various disciplines. The acquisition of new spectrum and subsequent technology deployment results in massive Capex and Opex.So simply bidding without an all encompassing strategic plan and its flawless execution is a recipe for disaster....even if your Uncle is running the Regulatory Authority !!!

-------------------------------♠-------------------------------

4.2 : Telco M+A in Africa : TDD means TOTAL DUE DILIGENCE

Nothing symbolizes the African renaissance better than the mobile phone. It represents technological advancement, deepening connectivity, and economic inclusion. Unfettered by outdated fixed-line infrastructure, Africa is at mobile technology's bleeding edge — pioneering everything from mobile payments to crowd-sourcing. Unfortunately political upheaval and commodity price volatility have posed a big challenge for investors in Africa . Furthermore Telcos are a capital intensive business so there are special ROI challenges in monetising in hyper competitive low ARPU markets that characterise Africa. In any case most Telcos must slim down their various network, sales and service business models, abandon secondary activities and use M+A or cooperative ventures to penetrate fast-growing markets.

Large telcos that operate in emerging markets have higher valuation multiples than their peers in mature markets. Investors place a premium on growth prospects, so long as they are well monitored and diversified . Equity investors we are looking for a business that generates positive free cash flow from wireless and mobile broadband operations by year 5, and a business valuation based on EBITDA positive operation by year 2 and a 5x EBITDA and 2x revenue multiplier valuation .Acquiring a profitable business with a large, established network may lead to future growth, saving time and money that would have been required to build a similar network from scratch. Acquiring a competitor may enable the acquirer to reduce price wars in certain geographical areas, or to protect itself against acquisition by a larger operator.

Capital markets prefer businesses that have a homogenous operational and capital structure because they are easier to understand and value. This is the reason that conglomerates often trade at a discount to the „sum of the parts" and end up being broken up. MNOs essentially have a highly capital intensive infrastructure-based portion of the business (deploying and running the network) and a low capex, innovation-focused services portion. Splitting these two parts could be perceived by investment bankers to be logical as it would enable management to focus, make performance more transparent, and valuation easier.

Cash Returns On Invested Capital (CROIC) is a good measure of company performance because it demonstrates how much cash investors get back on the money they deploy in a business. It removes measures that can be open to interpretation or manipulation such as earnings, depreciation or amortisation. Telcos tend to focus on the existing capital-intensive business (which currently generates CROIC of around 6% for most operators) rather than investing in new business model areas which yield higher returns.

The new business models (monetising Web 2.0 services) require relatively low levels of incremental capital investment so, although they generate lower EBITDA margins than existing services, they can generate substantial CROIC margins .You want to invest in a Telco with a consistent record of sales growth or a Greenfield that can deliver this in their business plan with a committed management team. Mergers in the telecom sector tend to build on existing "triple-play" offers. The emergence of "quadruple-play" offers — bundles of fixed telephony, broadband internet, mobile telephony and TV — are likely to lead to gains in market share and average ARPU, and a reduction in churn rates. There is a class of intelligent network investments that are relatively straightforward to implement and will yield a bigger bang for the buck :

- More efficient network configuration and provisioning
- Strengthen network security to cope with abuse and fraud
- Improve device management (and cooperation with handset manufacturers and content players) to reduce the impact of smartphone burden on the network
- Traffic shaping and DPI which underpins various smart services opportunities such as differentiated pricing based on QoS and Multicast and CDNs which are proven in the fixed world and likely to be equally beneficial in a video-dominated mobile one.

When evaluating wireless investments , the net profit margin is a critical metric : a fat margin means more money to expand operations, refresh network technologies and marketing and brand building . Invested capital is reduced through the deployment of more efficient technologies and processes that enable effective network capacity to be increased (including better network provisioning, traffic shaping, mobile Wi-Fi offload, femto/pico underlay network, network sharing, Multicast and CDN usage etc.Good acqusition targets are Telcos that can rigorously and swiftly farm out everything apart from their true core business can cut costs by as much as 12%, investment spending by 6% and the workforce by 50%. In return, they become more agile, engender a more entrepreneurial spirit and can realize the value of each part of the company.

Telecom-specific assets should be identified and valued using robust valuation techniques and methodologies. Tangible assets are generally valued by applying the cost approach since no prudent investor would pay for an asset more than the cost to recreate it or to reproduce an asset of similar utility (replacement or reproduction cost method) . Bear in mind that Networks are built and then adapted over many years challenge any operator's staff to keep accurate records of exactly what has been deployed. Poor record keeping affects the ability to audit physical infrastructure and connection topology, and also its management and maintenance.This is especially true when there are mergers and acquisitions; not only may there be many inconsistencies in each company's records, but there is also a significant challenge to integration of often disparate management and inventory platforms.

A simple Network assessment and inventory audit (as part of the DUE DILIGENCE process) can reveal various technical parameters in broad network domains such as : Infrastructure components and capabilities , infrastructure topology and traffic patterns , infrastructure design, capacity planning, and scalability , Infrastructure policy services , Infrastructure management and OAPM (Operation, Administration, Provisioning and Maintenance) , Business continuity policies and practices. So you pay for what exists in reality not in a cooked up asset register or a pompous mission statement.

According to Accenture to reflect the real value relationship, the bulk of the due diligence effort needs to focus on helping the acquirer understand the target's future prospects and how those fit with the acquirer's strategy. This can require a disproportionate emphasis on at least two due diligence work streams: strategic and operational.

Strategic due diligence involves validating the acquisition target's fit with the acquirer's strategic rationale for the acquisition, and understanding the target's market position and outlook to inform the price offered. Operational due diligence involves understanding the operational characteristics of the target (for instance, organization structure, IT systems, and culture) and hence the integration approach and timeline that will be required, as well as validating the target's operational and capital expenditure outlook to inform the price offered.

In contrast with the financial and legal varieties, high-quality strategic and operational due diligence do not generally require an army of specialist advisors. Rather, the due diligence work streams can be staffed by the acquirer's own people, selectively augmented with advisors who bring targeted insights, such as independent perspectives on the market outlook or an in-depth assessment of the timeline and cost to get to a common IT platforms.

Acquisitions and partnerships are essential for success in emerging market segments such as mobile advertising and cloud computing. However Telcos need to clearly discriminate between when they should acquire and when they should partner. The ability to sustain partnerships will emerge as a strategic differentiator. Effective management and implementation of M+A and partnerships offers significant operational upside to telecom players.

In the final analysis a careless approach to investment in Africa can be disastrous to the share price is evidenced by the Telkom SA / Multilinks Nigeria debacle. Over time about one billion dollars went down the drain in a classic case of poor INVESTMENT DUE DILIGENCE practices (or lack thereof).

---------------------------------♠---------------------------------

4.3 : Telco infrastructure sharing : Assessing benefits and overcoming challenges

At a high-level meeting during Mobile World Congress in Barcelona, senior leaders from eight major mobile operator groups, serving 551 million mobile connections across Africa and the Middle East, resolved to cooperate on network infrastructure sharing initiatives that recognise the profound impact of mobile broadband and Internet services on the citizens of both regions. The participating operators made this commitment in order to provide Internet and mobile broadband access to unserved rural communities and drive down the cost of mobile services for all sections of the population.

This initiative basically echoes the GSMA's call that telecom regulatory frameworks should encourage flexible commercial sharing arrangements and facilitate access to government-owned assets at preferential rates to help speed up the roll-out of new networks and support the business case to extend mobile networks into rural areas. Regulators should consider the competitive advantage that sharing of towers could provide in their respective markets. However, what they have to bear in mind is the fact that new and smaller operators will be incurring lease payments as an operating expense with relative lower risk, whilst the large and incumbent operators are still recovering the capital expense incurred in erecting the towers.

So what is really driving this network sharing phenomena apart from the altruistic motives of bridging the digital divide ?? Well how about the fact that increasing competition, along with investments in ever-changing technology, which has been pushing telecom operators towards new ways of maintaining margins. Since building and operating infrastructure is a significant cost for operators ,network sharing it is the ideal way to roll out infrastructure quickly and efficiently in low ARPU rural environments. operators can rely on a single set of infrastructure for their network. According to experts the estimated Capex savings resulting from tower sharing in the Middle East and Africa region amount to USD 10 billion. Quantifying and realising these savings requires a rigorous business plan and a meticulous execution controlled through appropriate contract governance structures and well-defined service level agreements.

Currently the most commonly shared infrastructure among operators is passive infrastructure, as it is easier to contract its set-up and maintenance. Sharing passive infrastructure only, means that newer operators still need to set up their own transceivers and other transmission equipment.Passive infrastructure sharing (commonly referred to as tower sharing) has attracted significant interest from both operators and tower companies. Companies like Helios Africa, American Towers and Eaton Telccom are already working to gain first-mover advantage by pursuing tower acquisitions in the region. Over the last 2 years, the tower business has grown into a fully-fledged industry in Africa and the Middle East.

Passive infrastructure sharing requires the consideration of many technical, practical and logistical factors although the principle is simple in theory. Any potential impact must be assessed and fully understood before sharing commences to ensure that there are no adverse effects on the operation of the site and the supporting network equipment and systems. Operators must consider items such as load bearing capacity of towers, azimuth angle of different service providers, tilt of the antenna, height of the antenna, before executing the agreement. Although, tower sharing enables new entrants to scale-up faster, it exposes established players to the risk of market share loss. Furthermore, the challenges of monitoring network performance and quality will increase as control over network roll out and equipment maintenance decreases.

As passive infrastructure business has evolved into a separate industry around the world, many tower companies in the telecom industry face several challenges. These include:

- High capital requirement: Tower deployment is a highly capital-intensive activity. The installation of each tower requires an investment of USD 55 000 to USD 75 000. Thus, tower companies the world over end up being highly leveraged

- Regulatory clearances: The first step should be to ensure that the regulatory authority is in favour of infrastructure sharing. Projects may stall because of delays in regulatory clearances. Apart from dealing with telecom regulators, tower companies also have to deal with other governmental bodies such as municipalities, forestry departments and environmental departments.Hurdles in obtaining clearance from a multitude of governmental bodies are often cited as reasons for delays in several site installations across developing nations. Since most of them are regional in nature, tower companies have to deal with quite a few governmental offices scattered across the country

- Operational cost optimisation: Although operational costs such as power and fuel are generally passed on to the operators, these are usually subject to agreed maximum limits. Thus, tower companies must work towards building controls to limit operational costs. Tower companies also face the problem of finalising the cost-sharing percentage and building a technology road map.

• Handling of local issues: Tower deployment and operation involves dealing with location-specific issues, including dealing with the landlord and local authorities, and running operations across a variety of geographies and terrains.According to KPMG , the accounting treatment for infrastructure arrangements would depend on the model applied and the structure of the transaction. Accounting for these arrangements could be complex and a detailed analysis of the substance of the arrangement is required.

Operators could: • Retain the infrastructure assets on their books (typically if risks and rewards of ownership are retained) • Derecognise the infrastructure assets (typically if risks and rewards of ownership are transferred to the third-party tower company) • Recognise a portion of the assets (typically if there is joint control over the asset)

In the US , TowerStream has formed Hetnets Tower Corporation, which will offer wireless carriers and others a range of shared infrastructure services and access for mobile wireless Internet services. They believe the explosion in mobile data in urban markets is driving a migration to small cell architecture, and the major carriers are presently focused on the densification of their networks. With the rise of mobile data placing a tremendous demand on the networks of the carriers, TowerStream concluded that its Wi-Fi network can serve carriers' data offload needs. To serve this need, Hetnets Tower Corporation rents space on street level rooftops for the installation of customer-owned small cells, which includes Wi-Fi antennas, DAS, and metro and pico cells. Channels on TowerStream's Wi-Fi network are available for rent for the offloading of mobile data.

Concurrently there is a growing industry in green technology that specialises in producing energy from renewable sources or with zero or reduced carbon impact. Such technologies include solar power, wind power, wave power and bio fuels. Operators should be in a position to benefit from these technologies as the amount of power they can generate continues to improve. Vendors have already successfully trialled a combined solar and wind powered base station in various African countries, which not only reduces the environmental impact of the site but also makes it more feasible for operators to deploy sites in remote regions by negating the need for traditional power supplies or maintaining a fuel generator.

Network roaming can be considered a form of infrastructure sharing although traffic from one operator's subscriber is actually being carried and routed on another operator's network. However, there are no requirements for any common network elements for this type of sharing to occur. As long as a roaming agreement between the two operators exists then roaming can take place. For this reason operators may not classify roaming as a form of sharing as it does not require any shared investment in infrastructure. When roaming agreements come to an end they can be renegotiated either with the existing host network or another operator with minimal effort and transitional impact.

Network sharing is increasingly favoured by progressive policy makers as a way of ensuring more rapid provision of 3G services and on environmental grounds. On this basis the European Union has consistently ruled in favour of permitting passive network sharing and more recently also national roaming under the caveat that competition rules are respected. The sharing of sites and masts, national roaming and RAN sharing tend to impact coverage, quality of service and pricing of services to consumers positively, as the cost saving characteristics of infrastructure sharing allow for increased efficiency.

One of the largest commercial infra sharing deals occurred in August 2004 between Telstra and Hutchinson in Australia This was cleared by the ACCC (Regulator) who assessed the benefits outweighed the potential competitive impact. Telstra paid $450 million to Hutchison Telecommunications Ltd for a 50% share in ownership and operation of its 3G radio access network infrastructure. The cost to Telsra of building a network over four years would have been $900 million to $1.0 billion. Telstra stated the deal was undertaken to save on costs of entering the 3G market and that they scored a tried and tested network at half the cost.Surely the same logic would apply to 4G LTE networks in MEA.

Ofcourse we have to wait and see in the MEA Telcos will actually realise the well known savings of passive sharing and whether some of these savings will be passed onto the rural consumers in form of lower data / voice prices. In spirit at least , infrastructure sharing is a step in the right direction for MEA telcos.

4.4 : Broadband business plans : mostly fiction and bamboozle

So why is the financial and investment community leery of business plans presented by broadband operators especially when it comes to broadband data ? To start with the Operator tries to bamboozle the financiers with tech speak " Guys we have an all-IP architecture, spectral efficiency OFDMA , bandwidth flexibility backboned on metro ethernet (Yawn) . Since the financers feel uncomfortable (if not downright ignorant) with the aforementioned techno blast they try to baffle the Operator with sublime finance speak " Guys , as equity investors we are looking for a business case that generates positive free cash flow from wireless and mobile broadband operations by year 5, and a business valuation based on EBITDA positive operation by year 2 and a 5x EBITDA and 2x revenue multiplier valuation" (Yawn).

After thoroughly confusing each other with their subject matter expertise both parties scramble to find some common ground : Clarion call " lets BRIDGE THE DIGITAL DIVIDE : meaning connect the poor sods in rural areas who don't even have water or electricity. Atleast thinking about the poor while we gorge on caviar sushi makes us feel human and that can't be bad nutrionally or spiritually that is !! As we all know the investment model for broadband wired and wireless installations must consider all aspects of design, deployment, and integration from the core through the systems architecture, service edge, access network and device. While the initial spend on deployment will have a large focus on capital components associated with procuring the necessary equipment throughout the network and systems architecture, as the broadband network service is introduced and subscriber adoption and usage rates grow, the ongoing operating expenses will consume a growing share of the total cost of ownership.

There are a lot of uncertainties connected to the forecasts. Since the broadband forecasts are developed thru qualitative and quantitative information, statistical modelling and also subjective input to the modelling, it is difficult to express the uncertainty by a pure statistical model. However, it is important to analyse the impact of the broadband forecasting uncertainty. The long-term forecasts are mainly used as input for rollout decisions of different broadband technologies and for establishing new network platforms. Techno-economic assessments are used to calculate net present value, internal rate of return and pay back period for the various projects. A relevant method for evaluating forecast uncertainty is to apply a risk analysis.

Typically the Operators grossly overestimate the market demand and thumb suck the number of new subscribers they will get in year 1 : guys we are targeting 5 % of the total available market without the foggiest clue what the real demand is. While statistical analysis and models are the tools of trade , the sane forecaster has to stand back and take a broader view. If a forecast seems implausible, this is generally because it is implausible. If a forecast results in an extremely high ROI, it is likely that the forecast underpinning the business plan is unrealistic. A very profitable industry attracts more competitors, leading to a loss in market share and increased price competition. This would change the firm's demand and revenue forecast.

A well researched and clearly structured methodology which is based on accepted economic theory and market models instills confidence in decision makers, investors and lenders. While forecasting subscribers and revenues for mobile broadband and specific applications is key to analyze the top line of any business case, translating these forecasts into traffic and bandwidth forecasts is required in order to effectively plan the network and analyze the impact on the bottom line. The moment you get the top line wrong then the whole excel spreadsheet is worthless.

To create accurate market-demand projections for broadband services and assess the availability of alternative technical infrastructures requires testing existing projections against different points of reference, adding context and detail where necessary. Where projections do not exist, we need to create them from scratch with management input. That entails conducting a macro-economic analyses to input to the overall expected level of spend on broadband data ; benchmarking with usage in comparable countries assists with the development of models of how that spend might be broken down by service type, and over time more advanced markets (such as Japan and South Korea) are analysed to assess how mobile data consumption might be expected to evolve over time.

It good to review brokers reports, market research and press articles relevant to the candidate market with care ; we must undertake interviews with relevant experts and panels in order to establish a consumption model for mobile data – where the consumer is likely to be, and what they are likely to be doing when they consume mobile data. To assess opportunities to achieve growth through the introduction of mobile broadband services, we must gather quantitative information and high-level financial information to develop a short business case along the following imperatives :

• Study the competitive landscape, the regulatory background and the alternative technologies available to develop a business model and a strategic positioning reflecting the brand, the skills and the ambition of the operator

• Develop a number of strategic scenarios, consistent with the operator's strategy, and propose models to assess the likely financial impact of pursuing each option

• Include the evaluation of opportunities such a diversification through organic or inorganic growth, and review of the synergies with the existing operator's activities and network

• Construct market demand projections for a comprehensive set of different content types .The output of this stage is a detailed spreadsheet model including projections of spend and usage for each different content type

• Once usage projections have been created, consider the 'optimum' technical support infrastructure for each service type. The result is a summary of the optimal technological choices in a given market, given the expected user profile

• Include the optimisation of returns and the minimisation of risks, as well as the identification and development of strategics for non-conventional revenue streams

Operators planning investment into broadband installations need to be certain that their front-end strategy and planning efforts consider the end-to-end proposition of the network, systems, and service to truly reap the cost benefits and the revenue potential of broadband wired and wireless services. And then and only then you might get lucky with the investment banking community.

Long-term broadband technology forecasting is not a very easy subject. Experience has shown that it is nearly impossible to make long-term forecasts without understanding the evolution of new broadband technologies and new broadband network platforms. Knowledge of broadband technologies regarding possibilities and limitations is important for the forecasting.In order to make good long-term broadband forecasts,techno-economic analysis of the relevant broadband technologies has to be performed.Each technology generates investments and operations and maintenance costs for the rollout, which is dependent on the characteristics of the various access areas in the countries.

The techno-economic calculations evaluate the "economic value", i.e. expressed by net present value or pay back period of rollout of different broadband technologies.The assessments are carried out for rollout on a national level and on specific areas like urban, suburban, rural and especially the rest market to examine the potential of the different broadband technologies. Therefore, the techno-economic analysis is crucial for technology rollout strategies and for broadband forecasts.

No matter whether an operator is established or new entrant, business modelling is used for strategic planning , project analysis and selection, internal monitoring and management, regulatory confrontation, etc. New operators investigate roll-out strategies for network infrastructures and use business modelling to present quantified economic cases to decide upon investment strategy, for lobbying regulators on interconnect agreements etc. Major public network operators use business modelling for tariff studies, interconnect pricing, cost analysis of services, competitor analysis and modelling, life-cycle costs of alternative technology strategies, business case evaluation and roll-out planning.

Business modeling is a combination of advanced and thorough economic methodologies and expertise and substantial and broad technology knowledge, including both technical and economic evaluation based on market, technical and economic parameters.

Business modelling methodologies have been developed in order to fulfil needs for business case development, making investment decisions or regulating telecom market.

---------------------------------------♠---------------------------------------

4.5 : Thinking Broadband economics : Key Insights Reloaded

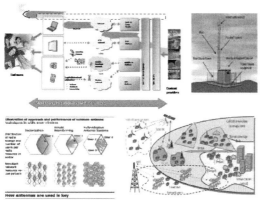

Mobile Broadband has seen feverish enthusiasm in the last two years, as evidenced by the number of LTE field trials taking place on a global basis. However full commercial deployments have been lagging because developing a coherent business case predicated on sound monetisation strategies remains an elusive and complex and process. The broadband market represents a tremendous opportunity, but there are definitely right and wrong ways to go about pursuing it.The industry needs an avant-garde approach that will leverage the inherent nature of an IP based architecture to accelerate the rollout out of profitable networks by paying specific attention to some critical elements which can make or break the LTE / 4G business case.

Today, communications service providers deliver traditional and IP services that span voice, data, video, content, prepaid and postpaid, fixed and mobile. To succeed, service providers must change to an infrastructure that supports new business models, real-time customer interactions, and new partners and channels.

Driving this transformation are underlying business applications such as billing and revenue management (BRM), customer relationship management (CRM), and enterprise resource planning (ERP). The new enterprise opportunity is a credible source or new revenue but there is still great confusion over where to start, how to scale and which divisions within telcos should be targeting these markets. Cloud Computing has captured the interest of enterprise customers because it offers flexibility and an ability to control costs. It fits well with the new Telco philosophy as it is about supplying assets as services.

When implementing billing and customer management platform several factors come into play : integration with existing systems, managing multiple billing systems, installation and service scheduling; trouble ticketing, billing blended services, managing channel relationships, automated provisioning, customer self-service, provisioning flexibility, and reporting. In fact, when properly integrated, a service based billing/CRM/Provisioning engine can evolve from an expense item to a revenue engine. If properly managed, efficient billing and CRM can translate into lowered costs, increased cash flow and increased profits.

As the telecom market becomes increasingly saturated, operators are looking for ways to stand out without resorting to price wars. Core services such as voice or data are becoming commodities; in order to avoid being limited to charging commoditized prices, operators must be able to create and deliver services that offer additional value, but are limited by the rigidity of many legacy service creation and OSS/BSS environments. A flexible policy management solution deployed either as a standalone implementation or as part of a larger OS S/BSS and service delivery transformation initiative, can enable value-added, differentiated services that can run on top of existing core services. While pricing can never be low enough from a consumer perspective, the ongoing quest for operators is to find a balance between competiveness and the ability to fund future investments.

In an all IP world , multi-core processors coupled with powerful virtualization technology enables the consolidation of all the physically discrete carrier-grade servers into a very attractive platform for low-end scalability. Replacing 20+ carrier-grade servers with either 2 blades or 2 carrier-grade servers based on multi-core processors represents a dramatic way to lower the cost of the core network elements required to serve the first subscriber; this type of radical consolidation represents at least a 10:1 reduction in initial CapEx, plus a comparable reduction in recurring operating expenses.

The greatest opportunity for revenue growth for wireless broadband presents itself in the form of smaller markets with less than 100,000 subscribers. By dramatically lowering the cost to serve the first subscriber using a consolidated core, new networks can be built on a campus or targeted community basis with new services tailored to the specific needs of these smaller, targeted markets.

Use of Smart Antenna technology is critical because Radio technology factors into any multivariate ROI equation in three ways:

- Range: dictates the number of base stations required to reach initial coverage objectives, driving the peak negative cash flow point
- Capacity determines how much revenue will be received per unit of capital expenditures on spectrum, equipment, and site acquisition, as well as per unit of operating expenditures on site leases and maintenance
- Coverage will affect marketing costs through its influence on unit subscriber acquisition expense and churn rates.

Any successful broadband operator will find their network making a transition from range-limited to capacity-limited site density much faster than typical voice networks. Simply because subscriber usage is 10x higher for broadband data and available spectrum options are limited. One critical element responsible for the wide variance in network capacity concerns how Operators use their antennas. Modern smart antenna systems (eg : MIMO) employ sophisticated techniques to control the distribution of radio energy in each sector Using average sector throughput as the measure, LTE can provide four times the throughput of 3G given the same amount of spectrum, but can scale to 10 to 15 times the throughput using a full 20 MHz of spectrum since LTE uses the most advanced antenna techniques.

Site sharing entails two or more license holders to acquire and rent a common site to host the Base Transceiver Station and transmission equipment. In addition, common equipment such as antenna systems, masts, cables, filters, outdoor shelter, SSC etc. can also be shared. Analysis has shown that sharing infrastructure will deliver healthy savings in capital expenditures (10% – 30%) and operational expenditures (20% – 40%) over a ten-year period. The reduced environmental impact is the green bonus for Telcos.

By cleverly combining the capabilities of a different technologies, the resulting broadband network architecture can provide high capacity for true broadband services (full triple play) and global coverage at an affordable cost. The various access, distribution and core network technologies in the hybrid recipe include:

• Wireless: Digital Terrestrial Television (DTT), Broadband Fixed Wireless Access (BFWA), Wireless Local Area Networks (WLAN), Free Space Optics (FSO), Satellite and Stratospheric platforms.
• Wire line: copper pair (for example xDSL), optical fibre, coaxial and power-line cable.

Mixing / bundling several technologies offers Flexiblility (several deployment scenarios and supported services can be fulfilled ; Scalability (for low-cost initial deployment) and Enhanced technical performance: in term of coverage, capacity and throughput.

Telco Execs must understand role of new generation applications, content and platforms to generate revenue from broadband networks assets as well as the critical role of billing engines and CRM systems and how to convert these into revenue generating assets. Success in broadband lies not just in deploying exciting new technology, but specifically in deploying a network and service creation infrastructure that is both cost effective, and capable of delivering – over time – the range of services that appeal to potential subscribers .

Operators evaluating broadband (wired and wireless) investments must extend their consideration beyond the accepted virtues of the technology and consider how the platform fits into their specific near and long term business model, measuring cost of ownership with potential for harnessing time-to-market advantages to grow subscriptions and generate revenue.

Telcos are being disrupted because the basis of competition in mobile has fundamentally changed. It has changed from "reliability and scale of networks" to "choice and flexibility of services", representing transition from "mobile telephony" to "mobile computing". Telcos need to move their innovation focus from technologies (be it HTML5, NFC, IMS, VoLTE, M2M or RCS-e) to ecosystems. That requires a much better understanding of how ecosystems are engineered, and how ecosystems absorb and amplify innovation. The change in the basis of competition is fundamental and irreversible.

Networks are facing a lack of scalable and sustainable architecture to meet the challenges ahead in terms of data traffic increases, video uploads and downloads, and enhanced M2M communication. Employing software-defined networking (SDN) techniques can help mobile carriers overcome those hurdles and attract new data-centric revenue streams.the promise of total cost of ownership reduction. Wireless carriers must aggressively jump on the NFV and SDN bandwagon, targeting integration across a multitude of areas including radio access network, mobile core, OSS/BSS, backhaul, and CPE/home environment

Telcos need to architect an easily managed network infrastructure that combines all the main elements : a consolidated network core to reduce the capex /opex , an efficient billing , PCRF and CRM model that is generates revenue , smart antenna technologies that optimize the spectral efficiency , mixing fibreoptic and satellite for cost efficient backhaul and site sharing to accelerate the time to market. In the final analysis the end-users will get a full set of broadband services at an affordable price and Telcos will not lose money through ill advised investments in technology.

4.6 : Carrier Billing is immediate revenue opportunity

Psst : If you don't like having your credit card tied to your Google Play account and live in Singapore, here's the good news: you can now link it to your monthly phone bill (or prepaid balance) from August 15 onwards if you're a SingTel subscriber. SingTel has announced a partnership with Google to be the first in Southeast Asia to offer carrier billing services. This means you can buy apps or in-app purchases and the amount will be charged to your mobile line instead . The company also mentioned that this will be rolled out to Optus in Australia, and companies where the company either has shares in or owns in Thailand , Philippines, India , Bangla Desh and Indonesia.

Aaah yes : welcome to the world of Direct Carrier Billing. Finally the Telcos have woken up to this latent and gigantic opportunity to capitalise on the App store phenomena. Telcos have always coveted a piece of the $770 billion mobile payments market. Direct Carrier Billing helps operators grow new revenue streams by leveraging their key asset: the billing relationship with consumers. Many consumers prefer carrier billing over credit and debit cards, as it simplifies the purchase experience and avoids the need to disclose card details to small or unknown merchants.

Content paid for via direct operator billing will come to a represent a $13 billion revenue opportunity by 2017, according to Juniper Research, compared to $2.3 billion in 2012. The research indicates that Google Play, Nokia Store, Blackberry App World and Windows Phone Store – all of which offer Direct Carrier Billing in a number of markets – accounted for approximately 48% of all app downloads in 2012. And for you NFC afficiandos : while trials have demonstrated extremely positive user responses to NFC– given the scale of the marketing/educational challenge facing MNOs and other NFC stakeholders – Direct Carrier Billing represents a greater and easier monetisation option in the short term !!

Direct Carrier Billing allows storefronts to enable payment amongst a far wider and diverse user base, both in developed and developing markets. In the latter case, bank account and credit card ownership is often extremely low; in the former, it provides a billing option to the prepaid sector and younger demographics. Furthermore, it enables few-click purchases, thereby making it a particularly attractive option for impulse purchases. Conversion rates are much higher than many other payment means, due to ease of : for example, the Google Play transactions processed on carrier billing in 2012 in the US grew by 3x in comparison with 2011. Nokia CEO stated that in cases where Nokia teamed with mobile operators to offer carrier billing, consumers were 5x more likely to complete an app store purchase than if the app store only offered credit card purchasing.

There is significant opportunity for Direct Carrier Billing to be utilised for 'real world' purchases in the lower value, higher volume area, both in terms of ad-hoc purchases (e.g. books, flowers) and more regular purchases (e.g. petrol). By adding ticketing applications and services to a mobile phone a customer could be less likely to replace their mobile operator with a new one; customer loyalty should increase as a result of mobile payments. With the addition of analytics of the BSS and Big Data, the operator can add far more value to mobile commerce. This is beneficial both from the perspective of the MNO – allowing it to personalise content discovery based on analysis of on-going individual consumer behaviour patterns – and for third parties such as advertisers and retailers. In several major cities in Central & Eastern Europe there are mobile ticketing schemes for trams and buses which are well established and see multiple millions of tickets sold annually. Examples include Prague, Warsaw, Bucharest, Bratislava, Zilina, Kosice and Tallinn

Telefonica Digital, along with the Telenor Group is now offering Carrier Billing to a combined customer base of 400 million. Impressive as that figure is – and i do not know how many of those customers have smartphones or tablets – the best figure is that revenues from payments have increased by 300 percent. Credit card and PayPal transactions have suffered as a result. In Latin America that figure is closer to 1,000 percent because of the low penetration of credit cards. In fact the new Firefox Marketplace, aimed at that market, is the first application to have Carrier Billing 'baked in.' Samsung has agreed to let customers on Telefónica's network charge their apps and content to their phone bill or take the payment out of prepaid credit, rather than having to use a credit card.

Typically new Carrier billing platforms are Cloud based gateway / middleware coded solutions dishing out SaaS . They enable Telcos to onboard and manage the largest set of app stores, merchants and aggregators, and process transactions for any type of goods – digital, remote, or physical. In addition, they make it easy to settle against any payment method (postpaid bill, pre-paid balance, wallet, or card) and quickly capture new and emerging revenue streams. These platforms include automated workflows and business processes that maximize operational efficiencies. While the cloud-based service delivery model offers low CAPEX requirements, the revenue-share pricing model on offer ensures that OPEX is tied to service providers' success and payments revenue.

We're going to be seeing more of these carrier billing arrangements in the future. Not only does it mean more apps and content will be sold, benefiting their developers, but it also means the Telcos themselves aren't shut out of the value chain. This shift towards giving Telcos a slice of the Apps Store pie is a good thing – not because the proactive operators deserve it because it rewards them for the investment in expensive network assets that enable the Apps and Mobile payment economy.

---------------------------------♠---------------------------------

4.7 : Future Track : Augmented Reality is cool and profitable

Augmented Reality (AR) is a hot topic right now, attracting much of the hype that was reserved for apps a few years ago. Is AR simply the next stage of development for existing value propositions, or will it bring with it entirely new propositions that offer new revenue streams for the Telecoms ecosystem ? Augmented reality relies heavily on video display and database transaction technologies that will necessitate significantly more bandwidth than can be delivered by today's 3G networks. When this is the case it is easy to see that the current mobile network infrastructure will not support broad rollouts of augmented reality applications, which can require multiple megabits of bandwidth as well as universal coverage.

Aahaa : enter 4G LTE with Digital Dividend Spectrum and you catch my drift !!!

In simple terms, 'Augmented Reality' applications and technologies bring users information that exists in the digital world and presents it automatically and intuitively in association with things in the real, or physical, world. Often, but not always, this information is from the web. AR is about creating, making explicit and displaying the relationships between the real and virtual worlds.

AR apps combine reality with a digital overlay allowing consumers to virtually try on glasses or items of clothing using their mobile phone. Usage of AR in the retail area can enable retailers to bring an internet-like experience into their stores, allowing consumers to see more information on a product simply by pointing their camera at it. With image-recognition transferred to the cloud, the number of images that can be identified will increase dramatically enabling retail brands to develop apps for use in-store. Usage by entertainment brands will also drive usage by driving consumers to download AR apps and try them out with products that they are familiar with.

AR has been studied in one dimension or another in labs for over 20 years. The most important drivers in AR uptake have been the release of new sensor-laden and processor rich smartphones with touch interfaces connected to cloud services by faster networks. In addition to enhancements in computing power, devices are increasingly pre-packed with GPS, compass, accelerometers and gyroscopes, adding to the now ubiquitous cameras and microphones. Thermometers, RFID and other wireless sensors are also appearing. At the same time a critical amount of information is available in digital format, meaning the potential for bringing the real and virtual worlds together through a mobile device is huge.

With the commercialization of high speed data network such as the Fourth Generation (4G) cellular networks via Long Term Evolution (LTE), the use of AR applications in healthcare represents a particularly compelling value proposition for cost reduction and of course saving lives. Many AR applications in healthcare provide the benefit of visualizing three dimensional data captured from non-invasive sensors. Applications range from remote 3D image analysis to advanced telesurgery.

Qualcomm Inc. launched augmented reality platform for Android smartphones last year. Further they claim that Vuforia, their augmented reality platform has seen tremendous growth with use in more than 1000 applications in the last one year. Google has also come out with Project Glass, which provides a way to search for information, read text messages and watch online video without having to fumble around with a handheld device. Intel has funded a augmented reality company recently. Nokia is coming up with city lens technology in its yet to be launched Lumia 920 smartphone which recognizes the current location of the user and guides him about the important landmarks using the phones camera.

Telcos are entering the market as application or software developers, or promoting the services developed by a third party. For example, Bouygues Telecom in France released the first in-house operator-developed mobile AR look up service in November 2009 with over 900,000 unique points of interest (POI), while Telefonica has a group in Barcelona R&D which is working on its own visual search technology. NTT DoCoMo offers its smartphone subscribers the intuitive navigation services "chokkan nabi" developed under contract for DoCoMo devices. Orange UK launched a free iPhone app and AR service for the Glastonbury Festival and others have released similar apps around special events.

Philippine telco giant Globe's Augmented Reality (AR) Christmas ad ran in major broadsheets with a colourful 5-part 'Parol' (Christmas lantern) depicting how Filipinos celebrate Christmas in diversity.The interactive ad allowed the audience to make their own Christmas "parol" and provided instant access for them to share it with their friends in social media. With AR, Globe hopes to establish a more personal engagement and intimate affiliations with brands, immersing them through different senses and forge a more robust interaction with its products and services.

Research firm Gartner Inc. proposes that augmented reality is one of the Top 10 strategic IT technologies of our time. According to research firm SEMICO, the total global revenue from augmented reality will touch 600 billion $ by the year 2016.They predict that more than 864 million mobile devices will be equipped with augmented reality by 2014 and more than 2.5 billion mobile augmented reality applications will be downloaded five years from now.

Science fiction or not that's big business : so clever Telcos will stake a claim to this new market NOW by partnering with AR app developers , Advertisers , Retailers to create and monetise new AR apps.

---------------------------------------♠---------------------------------------

Chapter 5 : Technology innovations drive Network growth

5.1 : Energy Saving in BTS : The impact of SON

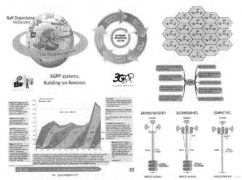

African MNO's are notorious (among many other failings) for their high dependency on diesel to fuel their base stations. One would think that faced with falling voice ARPU and hypothetical additional data revenue , energy expenditures would top the list to reduce OPEX since oil prices will remain stubbornly high at + $ 100 pb. Reduction in fuel OPEX requires CAPEX because it implies purchasing more energy efficient equipment or switching to renewable energy power solutions. While solar and wind remain the most prominent green technologies used to power off grid base stations , SON is another technical innovation within the 3GPP standards to save on BTS energy consumption.

So what is SON and how does it save on energy consumption ? A self-organizing Network (SON) is an automation technology designed to make the planning, configuration, management, optimization and healing of mobile radio access networks simpler and faster. SON functionality and behavior has been defined and specified in generally accepted mobile industry recommendations produced by organizations such as 3GPP and the NGMN . The first technology making use of SON features is LTE, but the technology has also been retro-fitted to older radio access technologies such as UMTS since Telcos began to understand that to meet the rising demand for data, it could be more cost-effective for them to expand HSPA and HSPA+ high-speed data capacity on the existing

3G infrastructure in many locations. SON promises enhancements in network efficiency, reductions in CAPEX and OPEX, improvements in customer experience (with potential reductions in churn).

So what are some of the benefits of SON in your BTS topology ?? A SON delivers an intelligent network where base stations self-optimize their operational algorithms and parameters in response to changes in network, traffic and environmental conditions. With operational intelligence at the access point, a SON can collect live network and call data, process it in real time, and either preview the changes or automatically deploy them live. SON offers offline planning capabilities for rapidly modeling the optimization of several parameters, including cell list additions, handover, interference control, and QoS enforcement. 3GPP Rel 11 has defined two energy saving states for a cell with respect to energy saving namely: not Energy Saving state and energy Saving state. When a cell is in an energy saving state it may need neighboring cells to pick up the load. However, a cell in energy Saving state cannot cause coverage holes or create undue load on the surrounding cells. All traffic on that cell is expected to be drained to other overlaid/umbrella cells before any cell moves to energy Saving state.

It is an indisputable fact that the traffic load in mobile networks is very unevenly distributed both over time and over cells. Excessive waste of energy occurs in low traffic situations since the radio system is optimized for maximum load. The NGMN Alliance has measured the daily average traffic distribution for an urban scenario .There is no communication activity in the cell in the two hours between approximately 04:00 and 06:00, and traffic is lower than 20 percent in the seven hours between approximately 0:00 and 07:00. So subscribers use more communication services during the day and very few in the wee hours of the morning. Knowing such usage habits is useful to determine how resources should be allocated so that the maximum amount of power can be saved. With SON base stations the network's coverage and capacity can be optimized when SON base stations can dynamically alter parameters such as antenna tilt and reference power offsets to compensate for lapses in coverage and ensure adequate capacity where it's needed.

Drastic improvements can be achieved by adapting to the actual traffic demand in a mobile network. The solutions include automatically switching off unnecessary cells, modifying the radio topology, and reducing the radiated power with methods such as bandwidth shrinking and cell micro-sleep. The challenge is to maintain reliable service coverage and quality of service (QoS) in the related area, while simultaneously consuming the lowest energy. The self organizing network (SON) supports proper selection of the appropriate energy saving mechanism and automatic collaborative reconfiguration of cell parameters with the neighbour cells.Mobility features like handoffs from one cell to the next can be optimized in a SON when base stations can balance load traffic among contiguous cells

Most engineers know that the amplifier power supply uses 60 percent to 80 percent of the energy consumed by base stations .If the RF power amplifier (PA) works at full power when there is no traffic or the load is very low, then power is wasted. Fortunately PA voltage can be dynamically adjusted according to the traffic load and required output power. When the output power is relatively low, the voltage required by a power amplifier is set lower than its maximum output voltage. In this way, power amplification is improved when the traffic load is light, and power consumption is reduced.In conventional amplifiers this power is independent of the amplifier input signal, i.e., of the current traffic load.

The key approach to saving energy is to make the power consumption proportional to the traffic load, either by implementing a partial shut down of amplifiers or by employing enhanced power amplifiers. During power-off of the amplifiers, further power savings can be achieved by also switching off the baseband signal processing, and indirectly, in the AC/DC power conversion and in the cooling fans.From the energy consumption point of view, low loads should be avoided. Instead, two types of mechanisms can be applied to reduce idle and unused capacities. As a first step, all energy-consuming equipment should implement power-reduction mechanisms while in operational mode, adapting to the actual load (short -term strategies). Second, the traffic should be reshuffled to a smaller number of highly loaded sectors or processing entities, and the others should be switched-off (long-term strategies). However, the coverage and the quality of service must not be degraded.

When there is no traffic, power can be saved by adjusting the PA voltage. Power can also be saved in transmitting LTE OFDM subcarriers. OFDM symbols can be automatically turned off when there is no baseband data transmission. This reduces power in the PA. Turning off PA OFDM symbols is more efficient than keeping them turned on all the time, especially when there is light or no traffic. PAs, cells, and power supply can also be intelligently turned off in the same way as OFDM symbols. A basestation's handling of RACH (random-access channel) offers another optimization metric. Automatically setting up a SON base station's RACH config parameters such as the number of preambles on a packet and ramp-up power can reduce synchronization times, call setup times, and handover delays while improving other aspects of RACH performance.

SON base stations are able to automatically configure themselves from the moment they are first powered up and before they join a wireless network. Once power is supplied, the base station would configure its physical cell identity, including its Internet Protocol (IP) address, and it would authenticate its software and configuration data. Following the completion of these baseline tasks, the SON base station would initialize the configuration of its radio by setting up its relationships with its neighboring cells and compiling its neighbor list. Based on a number of predetermined operational criteria such as energy savings, range requirements, and interference conditions, a SON base station will begin the self-optimization process once its initial configuration has been completed and it has joined the network. One of the first optimization tasks it will undertake will be to dynamically prune and select the base stations that are on its neighbors list.And this is where it really counts !!! A SON base station can save up to 40% of power usually consumed.

The combination of dynamic PA voltage adjustment and intelligent turning off of OFDM symbols is unique in the industry and can save about 32% of power consumption. Suppose the average power consumed by each base station is 1500 W (configured with three sectors). A single station can save up to 5200 kWh each year. This means more than 5.2 million kWh can be saved for a network with 1000 base stations each year, which is a saving of 1730 tons of standard coal and a reduction of 4500 tons of carbon dioxide a year. If base station power consumption is reduced, then less auxiliary power and heat-dissipation devices are required and less network OAM is necessary. The power needed for these devices is also reduced. New energy sources such as solar, wind, and bioenergy can be used in conjunction with these innovative energy-saving technologies. In this way, network energy consumption can be cut by + 50%.

The recent deployment of LTE to address the growing data capacity crunch, has highlighted the need and value of self-organizing capabilities within the network that permits reductions in operational expenses (OPEX) during deployment as well as during continuing operations. Self-optimizing capabilities in the network will lead to higher end user Quality of Experience (QoE) and reduced churn, thus allowing for overall improved network performance. Self-Organizing Networks (SON) improve network performance, but in no way replace the wireless industry's important need for more spectrum to meet the rising mobile data demands from subscribers !!

5.2 : M2M Reloaded : Crafting strategy to capitalise on M2M

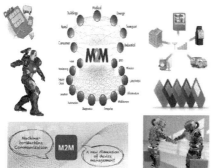

M2M is rapidly approaching a tipping point, a perfect storm of converging trends that creates the potential for fast and enormous growth. The democratization of device and service development is picking up momentum. You don't have to be an experienced engineer working for a top-notch technology company anymore to start building M2M solutions. The modularity that hardware and software developer tools introduce is lowering the barriers for developers and even end users to innovate.

The emergence of open, ubiquitous general purpose technologies will make it possible to develop, launch and maintain new applications with dramatically lower needs for capital and lead times. This would potentially be an open-source technical stack, analogous to the LAMP (Linux/Apache/MySQL and one of Perl, PHP, or Python) stack ubiquitous on the Web.

Analysts believe that Machine-to-machine (M2M) is often portrayed as a nascent industry sector. However, operators are already generating strong revenue in this sector, amounting to USD10 billion worldwide in 2013, and increasing to USD88 billion by 2023. Future growth opportunities will be realised in emerging regions as applications are tailored to local markets and the cost of solutions declines. All the market signs show that the industry is ripe for an M2M ecosystem to emerge in the coming years. Whoever takes the lead, whether it is telcos or other players, will have a strong advantage in building network effects and locking in users and developers. According to VisionMobile's M2M Recipe these are the key ingredients to the M2M banquet :

M2M is rapidly approaching a tipping point : Modular hardware and software components are lowering the barriers for developers and even end users to innovate, paving the way for new entrants in the market. As communications specialists, telcos are in pole position to take advantage of the unprecedented growth that innovation promises, but only if they play their cards right.

Ecosystem economics have proven to be a source of decisive competitive advantage in recent digital revolutions such as the PC, internet and smartphones. Most recently, Apple and Google used ecosystem principles to open up a whole new mobile computing market and created tremendous value in vibrant, large app ecosystems. Incumbents who missed the opportunity to leverage ecosystem economics have lost their market positions to the ecosystem driven newcomers.

It is only a matter of time before someone applies the same ecosystem principles to disrupt the M2M market. M2M is today where the app market was in 2008. Many of the roadblocks in pre-2008 app development can be found in M2M today: high market fragmentation, the lack of direct access to customers, complex, tightly controlled and expensive (for the service developer) distribution channels.

The source of possible explosive growth are new users, for whom current M2M solutions are too complex, expensive, or both. Making better, more advanced solutions for existing customers and system integrators will not attract those non-consumers. It is complexity and rigidity, not lack of performance that keeps new users from getting on board.

Non-consumption can best be addressed by leveraging a large and diverse set of developers. We don't know which M2M applications will emerge; indeed, it is fundamentally unpredictable. The app economy clearly showed that when you empower third party developers to experiment, they will find ways to create value, often in unexpected places.

The advantages of empowering external developers are clear. The likelihood of uncovering "killer apps" is greatly increased, the risk of failure is off-loaded to a large amount of companies and individuals and telcos can unlock a level of investments far greater than any single company could afford.

The owners of successful platforms connect users and developers. It is the network effects resulting from their interaction that will ultimately create unprecedented growth. A successful platform leverages three key control points: service creation, distribution and consumption. Each of these control points needs to be designed to reduce friction and amplify network effects. A platform strategy is only defensible if the platform owner can capture some the ecosystem's value. The trade-off between stimulating growth and capturing value can be managed by subsidizing or commoditizing ecosystem participants and by designing the platform to drive the telco core business.

Telcos have a strong incentive to own the M2M platform, rather than leave it to external companies. Only the platform owner can effectively use the tools of subsidies, competition stimulating openness and value-capturing closedness. Telcos can add value by making it easier to use their connectivity and providing more "M2M-friendly" interfaces – often described as managed connectivity. Beyond this, they can look to create and participate in the service enablers market for developers and application providers to easily identify, authenticate, provision, and maintain their device fleet;to update and rollback software on the devices and enable them to deploy processing logic into the "Internet of things" in order to render the system more robust, distributed, and autonomous.

AT&T's Digital Life initiative has packaged a set of home automation solutions around security, access, energy and water. This makes it easy for consumers to find and select solutions that suit their needs. On the one hand, because AT&T only sells products from carefully selected vendors, consumers can be assured of a certain quality standard. On the other hand, as AT&T manually selects solutions and presents them as its own, the company doesn't give the user much choice (tens of products at most) and makes it difficult for product developers to join the program.

The M2M Marketplace of Deutsche Telekom is an early example of what an app store equivalent for M2M might look like. Vodafone is able to provide a global SIM to support your M2M capabilities and applications across the globe. By providing a global SIM that is pre-provisioned and ready to use, they can significantly reduce the complexity of installation, distribution and deployment of M2M solutions.

Telcos that have assembled the appropriate teams and resources will be poised for greater success as the M2M market begins to grow. As operators in developed markets have learned, it takes 18 months or more to organise the various aspects of an M2M business.Telcos now have the opportunity to reap the first mover advantages they didn't seize as the smartphone disruption unfolded, to take up a position of control and to avoid the need to respond to a commoditization scenario. The time to act is now.

---♠---

5.3: The Diameter Storm in LTE : GET READY OR ELSE

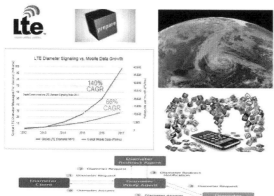

As 4G networks expand and customer adoption increases, the wireless industry faces a new "signaling storm" – this one related to core network signaling using the Diameter protocol. With the migration to 3G, LTE- and IMS-based networks, the need to manage the growth in Diameter signaling has become critical to optimize the network and ease congestion in real time. Operators are addressing concerns about data capacity by migrating to LTE, offloading traffic to Wi-Fi, and deploying small cells. However, the impact of network signaling has gone largely unreported.

So what is Diameter ?? Diameter is an authentication, authorization, and accounting protocol for computer networks. It evolved from and replaces the much less capable RADIUS protocol that preceded it. Diameter and SIP are the protocols in LTE networks that replace SS7, which is used in 3G networks. SIP is the call control protocol used to establish voice, messaging, and multimedia communication sessions. Diameter is used to exchange subscriber profile information and for authentication, charging, QoS, and mobility between network elements. The Diameter protocol was originally envisioned to handle things like charging or simple policy control but was expanded to take on a much broader set of responsibilities such as :

\# Registration, authorization, and authentication
\# QoS/bandwidth-based admission control
\# Charging
\# Location

The DSR (Diameter Signalling Router) is at the core of the new Diameter network. Like the central nervous system, which relays messages from the brain to different parts of the body, the DSR integrates with Diameter-based control elements, relaying messages among them. It handles traffic management, routing, and load balancing across Diameter elements to ensure scalability and manage congestion. Each endpoint needs just one connection to a DSR to gain access to all other Diameter destinations, eliminating the point-to-point Diameter mesh.

By moving intelligence from the network's edge to its core, the DSR improves signaling performance and interoperability between endpoints, streamlines routing, and reduces network cost and complexity. Network expansion is simplified, because routing updates, maintenance, and interoperability tests are centralized at the DSR. From its vantage point in the network, the DSR provides an ideal point for networkwide policy binding and protocol mediation. As the network demarcation point, it creates a roaming gateway; linkages to cloud, OTT, and M2M providers; and a central point for security and topology hiding.

Why the storm in the LTE cauldron ? LTE networks are characterized by more boxes and increasing complexity. This means that even more signaling is required for these boxes to communicate with each other. Carriers often deal with signaling at the level of individual network elements rather that at the network level. That breeds inefficiency and far too many point-to-point connections. As the signaling load grows, the resulting "n-squared" increase in core signaling traffic can quickly overwhelm nodes in the network. In January 2012 NTT DoCoMo networks suffered a 4 hour network outage due to an abnormal spike in signalling traffic. One recent network outage prompted the operator to ask Google to reduce the amount of signalling from Android devices.

The signalling storm has two fronts: radio frequency (RF) signaling hitting the radio access network (RAN) and surging Diameter signaling traffic in the core network. Although the end result in both is the same—network congestion, service degradation, and dropped calls—the nature and causes of each are very different. It is important for operators to distinguish between the two. One increases network costs, whereas the other drives significant revenues that more than offset the minimal cost. The importance of network architecture in overall performance, as well as its role in the success or failure in the mobile data market cannot be underestimated.Increased RF signaling fuels RF interference in the RAN, which decreases the amount of spectrum that is available for useful, revenue-generating flows such as user data. Moreover, RF signaling drives growth in other forms of signaling traffic, including S1-MME signaling between the eNodeB and MME and Diameter signaling in the network core.

Diameter is also the protocol for policy and charging in 3G networks. In both cases, Diameter enables revenue-generating personalized mobile data services, including tiers, loyalty programs, application-specific QoS, and value add for over-the-top (OTT) and machine-to-machine (M2M). There is a direct correlation between Diameter traffic and data revenue. As service providers begin to monetize their IP networks, the volume of

Diameter signaling increases. The Diameter signaling storm is being driven by the cumulative impact of the growth in connected devices and applications, personalized service plans, and an increasingly mobile subscriber base. The problem is compounded by the legacy Diameter architecture itself, which has no signaling core to efficiently manage the increasing load of signaling traffic passing back and forth between the Diameter-based nodes.

According to Acme Packet, by 2015, 44,000 Diameter transactions per second (TPS) will occur for every one million subscribers. For a moderately sized LTE deployment of five million subscribers, a mobile service provider will need to process between 220,000 and one million Diameter transactions per second. Online charging is the largest area of growth for Diameter signaling MPS as service providers migrate their charging networks to all IP-based architectures — forecast to increase from nearly 18,000 MPS to nearly 14 million MPS by 2016, for a CAGR of 280 percent. Subscriber authentication will generate 5.4 billion messages in 2016. Social networking activity is forecasted to spike from about 155 million Diameter messages in 2011 to 17 billion in 2016.

VoLTE and video streaming will sharply increase Diameter signaling with about 36 billion and 30 billion Diameter messages, respectively, in 2016 — generating high amounts of Diameter signaling due to Quality of Service (QoS) requirements. Policy has the largest impact on total Diameter signaling traffic with nearly 24 million MPS crossing LTE networks by 2016 in support of policy use cases —a CAGR of 269 percent over the 2011-2016 period — based on the growth of sophisticated data plans, personalized services, and over-the-top (OTT) and advertising models. The chattiest applications can generate as many as 2,400 network signaling events per hour. The increasing use of cloud-based software will add to the problem, because server-based data and applications must synchronize constantly with the device accessing them.

All of the resulting signaling will result in network outages because the servers involved in processing various AAA, QoS, or charging functions are not equipped to deal with spikes in volume. As proven during Hurricane Sandy and other natural disasters, congestion of the core signaling network is a key concern operators have to address when friends and families flood lines in search of loved ones. When the core fails, nothing works, therefore making the core becomes a critical component in the network. Geographic redundancy and traffic control is paramount to a robust signaling network that can survive any crisis.

It's time for operators to adopt strategies that allow them to manage traffic growth and mitigate the impact of Apps and connected devices on network signalling. Regulation of applications that require frequent updates may serve to ease signaling traffic to some extent, but it could stifle the popularity of the Apps that have become so appealing to mobile subscribers. Unlike the ongoing 3G signaling storm which is being created by smartphone-based applications and social networking services, the Diameter signaling phenomenon can also have positive implications in the form of new revenue-generating opportunities.

There is no slowdown in Diameter signaling growth in sight. Billions of connected devices taking advantage of new mobile broadband networks, new applications and new service plans are all conspiring to create an explosion in Diameter signaling traffic. The result? New stresses on operator policy servers, charging systems, subscriber databases, and gateways in IP networks.

In fact, the release of the Apple iPad 3 in September 2012 unleashed an unprecedented signaling tide on the network. More than 3 million iPads were sold in its first weekend on the market, and that has shown no letup in subsequent releases. Features such as a 5-megapixel camera, high-definition video, OTT services such as Facebook and Twitter, as well as a wide variety of data-heavy consumer and business applications are virtually guaranteed to spin off Diameter messages at a fast rate. Moreover, as sales increase, subscribers will begin using iPads outside of LTE coverage spots. Every time the device moves between LTE and 3G/4G access networks, it will have to register on the correct access network technology, which will generate yet another level of Diameter signaling.

Bottom Line : Only fatalists and fools don't see a Tsunami bearing down upon them at full speed !!

---------------------------------------♠---------------------------------------

5.4 : Inside Track : Carrier Aggregation in LTE

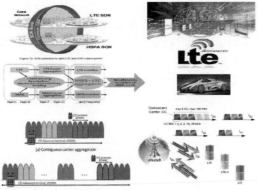 Carrier aggregation is one of the most distinct features of 4G systems including LTE-Advanced, which is being standardized in 3GPP as part of LTE Release 10. This feature allows scalable expansion of effective bandwidth delivered to a user terminal through concurrent utilization of radio resources across multiple carriers. These carriers may be of different bandwidths, and may be in the same or different bands to provide maximum flexibility in utilizing the scarce radio spectrum available to operators.

The key to achieving higher data rates with LTE is to enable network operators to use the technology in bandwidths wider than 20MHz. Some network operators may be lucky enough to have contiguous spectrum allocations of more than 20MHz. However, the nature of spectrum allocation over the years is such that most operators have a mix and match of spectrum within and between frequency bands. Following the redistribution of analogue TV spectrum and the provision of higher frequency spectrum, an operator might have LTE spectrum at one or more of 700MHz, 800MHz, 900MHz, 1800MHz, 1900MHz, 2100MHz, 2500MHz and 2600MHz.

Recently Telekom Austria, conducted a successful demonstration of LTE-Advanced carrier aggregation .The live demo, which included handling large file transfers with a simultaneous video stream, showcased download speeds of 580 megabits per second (Mbps), far more than twice the current 4G LTE peak rates. In combination with carrier aggregation, Telekom Austria plan to turn LTE-Advanced into a Gigabit technology that will allow a large number of users to simultaneously access high data rates within one mobile radio cell.

The U.S. market is one of the main drivers for deployment of Carrier Aggregation since the frequency spectrum is a scarce resource and very fragmented with few operators having contiguous 20 MHz spectrum generally available. The LTE Carrier Aggregation roll out will start during 2013 and we expect significant growth 2014. U.S., Korean and Japanese operators already have firm deployment plans for Carrier Aggregation. In the first phase, up to 20MHz of spectrum will be aggregated enabling subscribers to enjoy up to 150Mbps downlink data throughput, or even higher in the future.

CA (Carrier Aggregation) may be used in three different spectrum scenarios :

Intraband Contiguous CA — This is where a contiguous bandwidth wider than 20 MHz is used for CA. Although this may be a less likely scenario given frequency allocations today, it can be common when new spectrum bands like 3.5 GHz are allocated in the future in various parts of the world. The spacing between center frequencies of contiguously aggregated CCs (Component Carriers) is a multiple of 300 kHz to be compatible with the 100 kHz frequency raster of Release 8/9 and preserving orthogonality of the subcarriers with 15 kHz spacing.

Intraband Non-Contiguous CA — This is where multiple CCs belonging to the same band are used in a non-contiguous manner. This scenario can be expected in countries where spectrum allocation is non-contiguous within a single band, when the middle carriers are loaded with other users, or when network sharing is considered.

Interband Non-Contiguous CA — This is where multiple CCs belonging to different bands (e.g., 2 GHz and 800 MHz are aggregated). With this type of aggregation, mobility robustness can potentially be improved by exploiting different radio propagation characteristics of different bands. This form of CA may also require additional complexity in the radio frequency (RF) front-end of UE. In LTE Release 10, for the UL the focus is on intraband CA, due to difficulties in defining RF requirements for simultaneous transmission on multiple CCs with large frequency separation, considering realistic device linearity. For the DL, however, both intra and interband cases are considered in Release 10, while specific RF requirements are being developed.

Whereas GSM made use of only four frequency bands globally and WCDMA/HSPA requires five bands to get global coverage, LTE is deployed on more than ten bands already as of December 2012, and the number of bands will still increase. The frequency band fragmentation is a result of regional decisions on how the spectrum shall be utilized. Of course, there are initiatives to define some bands as global roaming bands. Another big challenge for Carrier Aggregation is to develop a RF front-end flexible enough to support a majority of domestic, and international, bands together with possible future Carrier Aggregation combinations

LTE Carrier Aggregation will provide mobile network operators with even greater scope to support services that hitherto would have been restricted to fixed networks, and may open up the possibility of providing a viable alternative to fixed network broadband services, particularly in rural locations where fixed broadband provision may be poor. Which is why CA is extremely important for Africa.

At the current rate of LTE Spectrum allocations it is most unlikely that the long suffering African broadband consumer will ever experience the high speed LT. By the time Regulators release the Digital Dividend spectrum in Africa (2015 - 2018), LTE Advanced will be a globally standardised and mature technology. This is the silver lining, believe it or not , since African Telcos will have the opportunity to launch LTE Advanced with all its speed benefits and refinements like Carrier Aggregation by the time they get some Spectrum.

---------------------------------♠---------------------------------

5.5 : Dumb Pipe Remedy : Transform the packet core to monetise Data

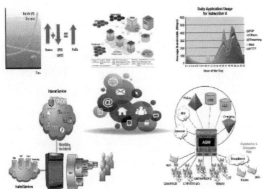

The availability of high bandwidth wireless packet technologies coupled with competitive flat rate pricing has fuelled an explosion in mobile broadband. Analysts predict that within 4 years 75% of mobile data traffic will be driven by video, but are Telco networks ready to cope? Are Telcos postioned to make money from this explosion in demand or will it be driving even revenues to the OTT players ? Peer-to-peer video download applications, for example, can quickly consume network resources and impact the performance of the overall network.

In the era of rapidly-increasing bandwidth demand but flat to decreasing ARPUs, service providers need smarter wireless networks to improve margins. These smarter networks can offer service providers greater control of their network to optimize operational costs and enable new revenue streams. Users are demanding a similar media-rich broadband internet experience in the wireless world as they have in the wired world.

The impact of mass market mobile broadband has a severe impact on the underlying wireless core network.It neccessitates the deployment of next generation of intelligent packet core technology. Referred to as the intelligent packet architecture mainly due to the flexible and flow based policy and charging mechanisms that are integrated into the GGSN these intelligent mechanisms leverage Deep Packet Inspection technology to enable services such as Content Based Billing and precise QoS and bearer resource management. Consolidating everything into a single intelligent packet core node provides for capex savings as well as a more coherent single point of awareness approach to delivering IP services.

This new generation architecture facilitates reliable, scalable, high density packet processing capability to meet the challenge of the data / video tsunami while reducing the overall cost of ownership for the operator The increased end user bandwidth available with mobile broadband technology is the resource for the sale of new revenue generating incremental services. These new services can be coupled with flexible flow based charging and DPI based content awareness to provide new and innovative marketing-led service offerings and charging schemes.

Understanding per-subscriber content traversing the wireless network is the critical first step to identifying problem areas as well as potential revenue opportunities. Which subscribers are consuming the greatest amount of bandwidth? Which applications have the biggest impact on your network? What are the time-of-day usage patterns? Fully understanding the usage of the network at a granular level is an invaluable tool to helping identify problem spots, causes of congestion, and which subscriber applications offer the greatest potential for profitability. Content management provides the answer to this business model challenge.

Content management platforms provide the means for operators to monetize the open Internet, make new revenues and, at the same time, optimize the cost of running these networks, making the service more profitable. It helps a network to transition from being a dumb bit-pipe to an intelligent profit-making network. In addition the use of an Intelligent ASN Gateway (Wimax) or S-GW (LTE) with its ability to manage bandwidth and deep understanding of real time content enables service providers to create unique and differentiated services.

Leveraging insightful reports and charts generated by the gateway management software, wireless operators will gain a deeper understanding of how their network is being used, where congestion problems may arise, and what revenue opportunities. This valuable information can be used for marketing, product planning or network control. Operators are able to extract a very granular view of content traversing the network on a per-subscriber, per-flow basis including hard-to-detect traffic, such as peer-to-peer file sharing or Skype, as well as standard applications such as email and video streaming.

The Gateway function supports extensive accounting and billing options for service providers, enabling them to effectively capitalize on service differentiation opportunities. In addition to traditional time and volume based charging the Gateway enables more flexible and intelligent charging solutions such as Content Based Billing (CBB), Flow Based Charging (FBC) as well as tiered and event based charging. Content Based Billing leverages the Deep Packet Inspection technology within the AGW / SGW while Flow Based Charging builds on the service data flow granularity provided by a 3GPP PCC compliant architecture.

The open Internet and rich mobile broadband model presents unique challenges to operators as they strive to effectively manage network resources and drive additional revenue and profit. Unfortunately , flat monthly rates with lowering ARPUs and the higher cost of burgeoning bandwidth consumption makes the mobile broadband business model challenging. Content management platforms help a network to transition from being a dumb bit-pipe to an intelligent profit-making network.

Rather than offering flat rate billing for opaque dumb bit pipes, operators can offer "smart" pipes that provide premium QoS for specific applications, such as video streaming or music downloads. The evolved packet core gateways leverage feature-rich, services at the edge architecture together with a full set of standardized and open interfaces. This delivers the key policy, charging and security enablers needed for service differentiation whilst protecting subscribers and networks from the threats posed by the new world of mass market mobile broadband.

---------------------------------------♠---------------------------------------

5.6: The Mobile VoIP Future is finally here

The emergence of voice over Internet protocol for mobile devices, or mobile VOIP, has the potential to transform the mobile device and telecommunications industries. The arrival of low latency LTE technology will finally resolve the latency problem in VOIP even as smart phones with integrated wifi continue to attract new customers to mobile service providers.

The capabilities of smart phones and a departure from the traditional contract structure for mobile wireless plans means that support for mobile VoIP is growing. The latest technology innovations such as low powered single chip radios with integrated VOIP protocol stacks, devices with integrated multimedia terminal adaptors and large scale, centralized Wi-Fi infrastructure promise to drive mass adoption of mobile VOIP.

The mobile VoIP market is expected to be worth $32.2 billion by 2013 and by 2019, moreover half of all mobile calls will be made over all-IP networks, according to recent industry reports. VoIP providers are also integrating Facebook into the desktop client. The social media giant has recognized the potential and is now offering its own mobile VoIP client for iOS. This move is just another indication of the role smartphones and tablets play in the overall mobile market, and the need for mobile VoIP providers to leverage strong partnerships with device providers.

For many enterprises , the optimum solution may be a combination of Wi-Fi and cellular. A Wi-Fi/cellular roaming solution requires dual-mode handsets that support both VoWiFi and cellular—and a network gateway. The gateway manages access and handoff and connects to both a mobile switching center for cellular calls and a data network for WLAN calls. As people move within range of a wireless access point, the gateway authorizes access and delivers both voice and data network services over the WLAN. When people move outside of coverage of the current WLAN, the gateway seamlessly switches control over to another WLAN or a cellular network if another WLAN is not available.

VoIP over a wireless LAN can provide easy internal calling for corporations, educational campuses, hospitals, hotels, government buildings, and multiple-tenant units such as dorms, with the ability to roam freely and advanced calling features such as voicemail and caller ID. Users can also use the LAN's Internet connection and an account with a VoIP provider to make calls outside the site, including domestic long distance and international calls, often at no extra charge.

We can now envision a scenario with interconnected fixed network environments and mobile network environments, in any combination, including 3G on the mobile side, and with fixed broadband, cable, and ISP environments on the other. When making the transition to IP, Telcos must keep in mind subscriber demand for seamless functionality and consistency across multiple service provider networks and types.

Standards based Mobile VoIP solutions offer simplified converged services between mobile and IP networks that reduce coverage disparities and operations costs to deliver competitive subscriber prices, consolidated billing, and subscriber loyalty. Mobile VoIP solutions utilize Network Convergence Gateways that leverage standards based on SS7 signalling enabling voice and data calls on any SIP-based client.

The coexistence of wireless heterogeneous networks has been widely recognized, and it has become more common that new mobile devices get equipped with multiple and heterogeneous wireless interfaces. Furthermore, the recent advances in software-defined and cognitive radio technologies including the availability of TV white space spectrum promise even more diversity and heterogeneity. This presents lot of opportunities and challenges for mobile wireless networking. Environment cognizance, spectrum-aware mobility management, and vertical handoff thus become critical components in the Mobile VOIP solution space as does correct network design.

The mobile industry developed a standard for all-IP operations called Internet Protocol Multimedia Subsystem (IMS). The IMS standard promises to allow service providers to manage a variety of services that will be delivered via IP over any network type – including the mobile network's packet switched domain (GPRS, 3G , 4 G). With IMS, service providers will use IP to transport both bearer traffic and Session Initiation Protocol (SIP) for signalling.

Any IMS strategy must include a solid plan for supporting subscription-based, usage-based and tiered billing. The ideal IMS solution must integrate smoothly with the OSS because the picture grows even more complex with content-based billing, context based billing and differentiated billing by QoS. Disappearing are the days of only flat-rate billing; transaction-based billing on application usage and subscription profiles is a likely future reality. You need an IMS solution that can reach across domains, for three big reasons:

1 : Quality of user experience — Your subscribers are going from one domain to another in real life. You need to provide a superior end-user experience and convenience wherever they are, on the same device or different devices.

2: Speed to revenue — When services can be delivered across access types, you'll see faster acceptance among a larger community of interest.

3 : Future-readiness — You need the flexibility to address new and evolved business models two to three years into the future, even if you don't know where your company will be.

Key Benefits of Mobile VOIP include but not limited to :

• Communications service providers can capture wireline Minutes of Use (MOU)
• Improves quality of mobile voice and data services in homes, offices and public access venues
• Extends the subscribers single identity to the fixed-wireline venues and strengthens the brand to create a "sticky" service
• Open architecture yields best performance at lowest infrastructure cost
• Extends mobile footprint via unlicensed spectrum thus reducing capital expenditures on mobile base station construction and

infrastructure
- Leverages the economics and prevalence of Wi-Fi and SIP
- Maximized existing and future infrastructure investments by deploying pre-IMS (IP Multimedia Subsystem) network elements today.

O2 UK is launching its Tu Go app the operator said will enable its users to make and receive calls, texts and voicemail via the Internet using their existing telephone number.The service, available on all Apple and Android devices, is free to download for O2 contract customers, with the calls and texts taken from their existing bundle.The aim of Tu Go is to free customers from being locked to a single handset : Customers can now take their mobile number wherever they like, even away from their mobiles !! O2 UK customers can be logged into the Tu Go service on up to five devices at once, said the operator.

Incoming calls will ring all logged-in devices, including handsets using SIM cards associated with different networks and Internet-enabled gadgets such as iPods. Only available to O2 UK's postpaid customers, it is a cloud-based telephony service, allowing the user to register multiple devices and make and receive calls and messages from all of these as if from their telephone number. Any usage comes from the user's postpaid inclusive bundle.

TuGo can therefore be used regardless of physical location over Wi-Fi using the user's home contract. This also makes it an FMC solution, because it will work indoors on Wi-Fi at places where mobile coverage is poor.Significantly, the integration with native communications services (telephony and SMS) means that users are not restricted to communicating with other TuGo customers. Exchanging calls and messages with users of basic services works well, with information (e.g. caller ID and dialed number) shared between the native dialer app and the TuGo service. All communication is organized in threaded timelines, and is displayed regardless of the device used.

Facebook paid a whopping $19 billion to acquire WhatsApp. Currently a messaging application, WhatsApp is looking to implement voice over internet protocol (VoIP) by June of this year. A prime example of the growth and prosperity of VoIP in the business market today, WhatsApp already has 465 million users – all of whom will be able to make VoIP calls after WhatsApp releases an update by the second quarter of 2014.

Africa is poised for Mobile VOIP revolution with the arrival of ample submarine bandwidth , the continued expansion of terrestrial fibre optic , low cost dual mode smartphones on prepaid and budding LTE networks. So the big question is : Are the Telcos going to profit from Mobile VOIP by offering carrier grade solutions to consumers and enterprises or will they let the OTT players to pile in and eat their lunch as usual ??? Time will tell

---------------------------------------♠---------------------------------------

5.7 : Telco CEM : Implications in the Digital Era

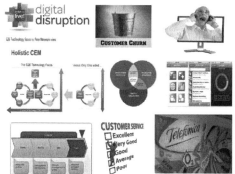

How should Telcos should manage the customer experience parameters in the digital age ? According to a recent Aspect Software study, almost 75 percent of consumers prefer to solve issues on their own—almost one out of three respondents noted they would rather talk to a toilet than a customer service representative. Mobile operators suffer from some of the worst levels of customer satisfaction in the world eventhough all the tools and platforms to offer superlative experience are available to them. What a disgrace !!!

The annual WDS Loyalty Audit has revealed that ONLY 35 percent of customers are highly satisfied with the mobile operator. More worrying for carriers is that a quarter of subscribers claim low satisfaction.The figures pose serious questions for operators, as feeling satisfied is intertwined with a customer's intent to repurchase. A unsatisfied customer is 8 times more likely to switch operators. Despite these disappointing figures, mobile operators still appear to be doing very little to understand their relationship with consumers or customer loyalty. This in turn means their loyalty programs and customer satisfaction schemes, vital to customer retention and solid business performance, remain outdated and fail to deliver.

Customer Experience Management is certainly one of the biggest current buzz words in the mobile arena. A lot has been said and written around CEM but still there is no unique industry consensus about what CEM actually includes. From managing the brand's perception across websites or street-shops to subscribers' actual appreciation about the services and applications they pay for, Customer Experience Management is claimed to be a part of every business process. A common misconception in the industry is that CEM is a replacement for CRM which simply is not correct.

As the industry moves into a growing market of digital services built on infrastructures that enable fast development and deployment of new services, the service portfolio itself is not sufficient to establish a lasting differential in the market place. Such a differential is quickly eroded by competitive service providers. Having tried to differentiate through technology and 'clever' pricing models and found the strategy to be short lived, service providers are realising that a more solid differentiation can be gained through managing the customer experience. This does not just mean delivering service that meets the customers' expectations but that all aspects of its business must support the concept of a superior customer experience.

Many countries have a penetration rate above 100%. In such a competitive market, churn has become the major concern of operators who have changed their priorities from customer acquisition to customer retention. Operators' financial reports show that, for a medium sized operator the average cost of retaining an existing subscriber amounts to tens of dollars per subscriber per year. Comparing this with "Cost Of Acquisition" for new subscribers (COA), often worth several hundreds of dollars per new subscriber, operators are inclined to pamper their existing subscribers, especially those generating higher Average Revenue Per User (ARPU). In this context it is key for operators to understand user expectations and adapt mobile data plans to their needs.

A successful transformation into the CEM world can only be achieved by building on top of good CRM processes and practices. CEM takes us a step closer to achieving improved customer satisfaction. Instead of asking the question, "This is what we are doing, how well are we doing?" which is a CRM approach, CEM asks, "What is important to you, and how well are we doing?. CEM is aimed at turning customers into fans by seeing the world through their own eyes. In the May 2012 issue of Telecom Buzz, published by MobileComm, the article "Customer Experience Management: The Next 'Buzzword'" declares the four pillars of telecom CEM ...no.... not nuclear physics or Astroflight.... but simply :

- network experience (includes coverage, signal quality, speed)
- commercial experience (includes billing, payment)
- product experience (includes telecom products such as handsets, VAS)
- service experience (includes after-sales service , customer queries)

In the social networking age you would already know the crucial role of social media and mobile marketing are the new building blocks when developing your future-proof CEM strategy. Increasingly Telcos are looking to looking to the social networking sites to provide valuable feedback on what the customer is experiencing. Twitter and Facebook provide rapid indicators on when things are going wrong. Systems that automatically monitor key social networking sites must be deployed to flag to the Service Management Centre when traffic increases. Often this is the first sign that a service is failing or a new service does not work the way that it should.

Delivering an effective Customer Experience Management requires a coordinated program across the entire organisation and is best achieved by adopting a maturity framework similar to the Capability Maturity Model Integration (CMMI) framework.

CMMI is a proven process improvement approach whose goal is to help organizations improve their performance. The TM Forum is developing a maturity model for the implementation of CEM. This CEM model, as with CMMI, is a five stage model that guides the service provider on a journey to a fully implemented and controlled CEM environment.

The measurement of Customer Experience is based on measuring the extent to which the customer's needs are satisfied using customer/user centric measures such as: + Would advocate (e.g. churn and loyalty indicators) + Would recommend (e.g. Net Promoter Score) + Would Buy again + Product availability + Product usability.Having the right tools and OSS / BSS environments in place to support CEM is absolutely critical to achieving the end goal. As such establishing an early dialogue with tools suppliers (internal and external) has to be a priority in the early days of the program if for no other reason than the lead times for delivering and integrating the necessary solutions.

Bear in mind having CEM people without equipping them with supporting tools will lead to frustration and will feed the 'naysayers' with ammunition to criticize or undermine the CEM program. CEM is likely to introduce new working practices which may to some, seem unnecessary and a hindrance to rolling out new digital services quickly. Without strong governance the program will become disjointed with different parts of the organisation going their own way instead of a single 'joined up' approach to delivering a good customer experience.

Network management plays a major role in an improved CEM. By avoiding network congestion and poor performance, telecom operators improve the quality-of-service (QoS) level, which eventually can reduce churn to a great extent and increase customer satisfaction with an operator. However, an improved network QoS represents only one of the factors that influence the overall "customer experience" equation. Strategies for reducing churn need to take place at every step of the customer life cycle. If marketing communicates something that is not supported by product quality, network infrastructure, billing processes, or customer care/service teams, the relationship worsens until, ultimately, the customer moves away.

Recently Telefonica implemented a suite of CEM tools and platforms to improve the end-to-end customer experience across mobile data, mobile voice, IPTV, high-speed Internet, cable, satellite and voice services. The platforms and tools enable their customers to troubleshoot and manage their digital experiences through devices such as mobile phones, laptops and IP set-top boxes, via dedicated web portal and apps.Telefonica's approach towards managing the customer experience embraces a multitude of critical success factors including customer surveys, social media activity, contact centre stats and service specific data. Recognising that the CEM view is a complex but profitable undertaking if you get it right , forward looking Telcos such as Telefonica have developed an OSS/BSS environment that enables them to display disparate customer data in one single 'vital signs' view.From this single view Telefonica are able to calculate various Customer Satisfaction Index values which they can then use to drive their customer centric quality improvement programs.

For the telcos to remain competitive an overarching customer-experience strategy ultimately makes more business sense.With growing pools of data (both structured and unstructured), gaining individual customer insights and coming up with products and services that suit them would require a sophisticated level of business intelligence such as CEM, which would deliver performance analytics to all management levels. Telcos that continue to ignore the dire need of having an effective CEM strategy supported by appropriate tool sets might eventually find themselves thrown out of the race...so good riddance and the Telco industry will be all the better for it !!

5.8 : Getting to grips with Telco API business and Developer Community

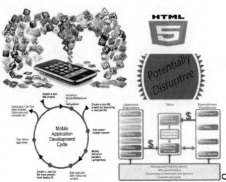

Since 2008, mobile software and applications have moved from the sphere of cryptic engineering lingo to vital part of the essential marketing playbook for mobile industry vendors. In stock market terms, developer mindshare is one of the hottest "commodities" in the mobile business, one whose "stock price" has ballooned in the last few years. Platform vendors, handset OEMs, network operators, hardware vendors, and infrastructure providers all want to contribute to mobile apps innovation. All the Telco value chain players are now vying to win software developer mindshare, in order to add value on top of their devices and networks. Mobile application development and integration are at the forefront of the modern SOA story.

Key mobile trends in 2012 included the emergence of app stores, the HTML5-native debate, mobile back ends, RESTful services and open APIs. Widespread adoption of mobile applications is at the root of these changes in development, and it shows no signs of decline with the consumerization of IT. In general, mobile web development within an HTML5 browser or web runtime is promising when it comes to market penetration, ease-of-use and cross platform support. Industry analyst Gartner Inc. predicts that more than 73 billion mobile applications will be downloaded in 2013, and that that number will nearly quadruple to 287.9 billion by 2016. Cloud mobile back-end services stand to become a key component of the application development ecosystem. By 2015, 80% of all mobile applications developed will be hybrid or mobile-Web-oriented .But how is the landscape of mobile developer mindshare looking today?

The recent report " Mobile Developer Economics & " contains new insights into the motivations of mobile developers . For instance , developers still consider fun and coding speed as very important even if developer mindshare is turning towards the appeal of monetization and reaching a large audience. The technical reasons for selecting a platform seem to be gradually becoming a less important selection criterion. For today's mobile developer, market penetration and revenue potential are the two most important reasons for selecting a platform. The most successful developers are those that extend apps to new markets, either to new geographies or different verticals. To some extent, these strategies rely on copying the recipe of an already established and successful business: these are apps that have been tried and proven in at least one market and are generally less risky options for developers.

Application developers are also increasing their demand for app store platforms, which provide a centralized place to buy, sell and manage their apps. More applications are designed with open APIs to enable application-to-application integration. As a result, the relationship between business and developer is shifting to give external developers more sway. However, the study shows several pain-points with mobile web technologies compared to native applications, namely issues with development environments, device API support and UI creation.

The goal of Telco API programs is to allow developers to take Telco services into new niches and use cases, and scale from hundreds to thousands of partners. Some of these new use cases will result in supplemental Telco revenue streams, some will facilitate customer acquisition, while others will subsidise ecosystem creation costs. APIs need the flexibility to allow developers to experiment with new use cases, and thus discover and satisfy unmet user needs. If Telcos allow and encourage developers to create locally-relevant differentiation on behalf of their subscribers, their fragmentation disadvantage could transform into the advantage of local presence.

Since there is no such thing as an "average developer", Telco API business models need to be designed to target one or more specific developer segments. To reduce friction and help developers discover new user needs and opportunities, Telco API business models need to subsidize experimentation and be designed for the ability to fail and retry cheaply. More specifically, if developers are charged based on Telco API usage, the app's business model must have a stable, usage-based income stream. By allowing free, small-scale usage of the API, Telcos permit developers to experiment with multiple business models, including free, until a sustainable, workable business model can be found.

Telco APIs will always be at a disadvantage versus players with global reach, if positioned in direct competition to native platforms or Internet companies. To be successful in API initiatives, Telcos need to consider developers as value-added resellers, and therefore design their API propositions for win-win outcomes. In other words, the business models of Telco APIs need to be aligned with the business models of developers. It is important to note that the same ecosystem economics that work for Telco APIs and app developers can be applied to other types of partners and service providers, such as Mobile Virtual Network Operators (MVNO) or machine to-machine (M2M) initiatives.

MVNOs can build ecosystems around the distribution business layer. App developers can build ecosystems around the service layer. And M2M companies, meanwhile, can build ecosystems around the connectivity business.

But be warned : The Mobile App economy displays one of the key tenets of The Disruptive Technologies Model : which postulates that.... the pace of technological progress generated by established players inevitably outstrips customers' ability to absorb it, creating opportunity for up-starts to displace. This new theory provides a useful gauge for measuring not only where competition will arise but also where, in an industry's shifting value chain, the money will be made in the future. In the 4G world the dominant firm-level MNO value chain is ripe for unbundling in response to accelerated product/services evolution. Future success within the industry will go only to those Telco players with strategic foresight to "skate to where the money will be" : solo or via partnerships !!!

Chapter 6 : New services to delight the Digital Consumer

6.1 : FREE FRANCE : Freaky and Ferocious

If you were to ask execs of France's incumbent Telcos who they loathe publicly and admire secretly they will probably mumble FREE and Xavier Niel. Now how did Free and Niel freak out fat cat telco execs who were relaxing on their oligopolistic sofa ??

In April 1999, Free entered the Internet service provider (ISP) market with a simple, no-subscription service. This commercial strategy was at first based solely on providing "Pay-as-you-go" access and enabled Free to win a large share of the dial-up market with relatively small advertising outlay as compared to its competitors. After completing the rollout of its telecommunications network and interconnecting with the France Telecom network in April 2001, Free was in a position to control the cost structure of an offering based on Internet connection time.

The Illiad Groups's network (Free is the brand name) enables it to design sustainable service offerings that are easy to understand, technically sophisticated and attractively priced. Free has capitalized on the different nuances of its brand name, transforming it from a name implying that the offering is free of charge into a name associated with high-quality paid services and the freedom offered to users of these services.

The high speed broadband Internet access offerings are among the most competitively priced on the market in their respective segments while providing a high quality of service. This positioning is a central factor in the Group's strategy and is aimed at creating the right environment for lasting and profitable growth . Through the use of its network and by building on its experience in dial-up offerings, Free has developed a high-quality broadband access offering which is attractively priced and, where possible, makes the most of the opportunities afforded by the unbundling of the local loop. Today French broadband rates are among the lowest in the world.

One of Free's most powerful offerings is the modem "Freebox". Rejecting the offerings from established equipment manufacturers, the Iliad group(who own the FREE brand) built its own set-top box running Linux to deliver Internet, voice, and TV services and designed its own DSLAMs to direct traffic into and out of subscribers' homes. This router is lent to customer that provides every services offered by Free, wifi access point, telephone connection to use VoIP service, and television signal access point. Free has always offered its services as a complete package with only one price 29.99euro. Last year, the company evaluated its network in anticipation of continued growth of subscribers and the planned service offering. Within the access network, FREE recognized the need to increase capacity, because IPTV and other high-bandwidth services were pushing the access network to its limits. Rolling out its own fiber local loop allows FREE to operate more independently from France Telecom, while improving margins and strengthening its service differentiation in the marketplace and is designed to accommodate multiple Service Providers, optimizing its return on investment (ROI).

The Illiad Group's and FREE success is highly dependent on maintaining its relationship with Xavier Niel, Senior Vice-President of Iliad and the Group's majority shareholder. Xavier Nel is sometimes called the " Steve Jobs " of France so you know the kind of mercurial character we arc dealing with here. In 2001 the iconoclastic businessman—who got his start running an online sex-chat service—launched the country's first triple-play Internet, television, and telephone service. He quickly grabbed one-fourth of the market by setting prices far below those charged by industry leaders, including France Télécom.

After thoroughly disrupting the ISP market by introducing super-cheap Internet access Xavier Niel pulled it off again with its mobile offering based on a 3 G license (2.1 GHz and 900 MHz bands.) With rates as low as €2 (about $2.60) a month, the Free Mobile phone service has lured an estimated 2 million subscribers away from competitors. Free upset the French mobile landscape by using very clever technology, marketing and financial tricks. As a company with a hacker culture, Free is a good example of how to execute against well-established competitors. Free has created an offering that you cannot ignore. Imagine a mobile phone plan, with unlimited talk, unlimited SMS and MMS messages, tethering and, even more important, unlimited data with a speed reduction after 3 GB. Usually for that plan in the U.S., you would pay more than $100 for limited data with a two-year contract. In France, it costs $25 per month and there is no contract. Competitors had no choice but to lower their prices, even if it meant lower margins and lower infrastructure investments.

Before launching the mobile offering Free revamped its Internet offering by bundling unlimited calls to mobile phones. It allowed Free to increase the price to $45 per month (€35.99) and therefore greatly improve its margins for its millions of customers as well as invest in its mobile future thanks to its triple-play margins. Free signed a roaming deal with Orange so they could launch the mobile offering quickly and without doing immoderate infrastructure investments. Another unexpected advantage is that Free had the largest hotspot network in the world because according to them every triple-play modem) FREE BOX) was in fact a hotspot. Thanks to the EAP-SIM protocol, smartphones could connect seamlessly to those hotspots

"There isn't another operator in the world that has won as many subscribers as us in such a short time," boasted Niel at a recent press conference. Free's foray into mobile has already shaken the country's top three operators—France Télécom's Orange, Vivendi's SFR, and Bouygues Télécom—which until now controlled 90 percent of the market and charged the highest rates in Europe. France's national telecommunications regulator recently warned that competition from Free could force the incumbents to shed as many as 10,000 jobs. Orange and SFR have each reported losing about 200,000 subscribers to Free.

---------------------------------♠---------------------------------

6.2 : Telco Accelerators : Hitting the digital tarmac at top speed

The latest issue of Ey Inside Telecoms 15 has a very interesting update on Telco Accelerator Programs. Now what is that ?? Basically when Telco operators establish new incubators as part of digital transformation initiatives with the creation of new business units to nurture entrepreneurial start ups. Operators are working with partners as part of accelerator initiatives, underlining how service innovation is increasingly reliant on harnessing expertise from across the telco ecosystem. Over the last few corporations have moved to an open innovation model or outsourced R&D. They're doing less basic research in house and essentially looking to bring that in through acquisitions. Ask Google and Facebook what they did to become so big !!

So why the sudden surge investment in Accelerators by Telcos ? For one the appearance of incubators in emerging markets underlines the role operators can play as leaders of technology ecosystems, particularly in countries where entrepreneurs may lack support structures. Accelerator programs offer a number of established advantages to operators — from lowering the risks and costs associated with new projects to enabling operators to tap into a wider talent pool. Not bad at all because it is not a merely a CSR initiative but a genuine effort to create new digital products and services with the help of small time developers on a shared risk basis : you know ...the same kind of small timers who built Facebook , Instagram , Twitter , WhatsApp etc into multi billion dollar titans that threaten to render Telco networks into unprofitable dumb pipes !! With the cost of developing new technologies coming down so dramatically, it makes sense for corporations to take smaller bets on new technology offerings developed in partnership with entrepreneurial apps developers .

As an example Wayra is Telefonica's Accelerator program. With the financial backing of Telefónica, and with the support of a global network of mentors, investors and partners, they aim to help the best entrepreneurs to grow and build successful businesses.They know the next digital revolution can emerge from anywhere, in technology nobody has the final word. If you have a business or an idea that uses technology to solve the problems of the future, Wayra is your place to make it grow so they opine. Their irresistible value proposition is : Our acceleration programme will give you funding up to $50,000, an incredible place to work, mentors, business partners, access to a global network of talent and the opportunity to reach millions of Telefónica customers.

In a typical deal in exchange for an equity stake of around 10 percent, successful applicants will gain access to an office space in the newly established Wayra Academy, financing, mentoring, access to technology expertise within Telefónica. After six months in the programme, they will be turfed out of their luxury abode and released to fend for themselves in the real world. Wayra is aimed at early stage technology startups less than two years old working in spaces including cloud services, financial services, M2M, digital security, e-health, mobile applications, social networks and e-learning.

A truly disruptive hardware or core technology investment can generate some lofty returns for Telcos who are battling with stagnating returns in the 4 G data era.Marrying incubation activities with a wider program of crowdsourcing, ecosystem collaboration and venture capital funding in an effective way is no easy task,and regular reviews of innovation agendas are a must. While many operators are focused on partnering frameworks and incubating new services as a route to service innovation, a robust internal collaboration structure should never be overlooked.

The « Orange Fab » is open to all US-based start-ups and is managed by Orange Silicon Valley, Orange's development center in San Francisco, California. Orange Fab aims to help start-ups grow their businesses and develop amazing products and services that benefit Orange and its customers. There were 4 – 6 start-ups selected to be part of the program that started in May 2013. Start-ups joining the program have access to mentoring sessions from notable Silicon Valley entrepreneurs, world-class engineers and experienced designers. All the start-ups that are selected for the program were offered $20,000 in funding.

The French operator has expanded the Orange Fab program to markets including France, Japan and Poland, with additional innovation investments in the likes of India and Tunisia.17 Africa is proving to be an important focus area: in May, Orange set up an incubator called CIPMEN to support SMEs in Niger, having already set up incubators in Senegal and Mauritius. Meanwhile, Millicom has set up a tech incubator in Rwanda, launching a purpose-built facility in the capital, Kigali, in March.

Believe it or not one of the pioneering Accelerator programs (MAGNET) was initiated by Motorola over a decade ago. The

MAGNet program provided application developers, included a membership structure, providing to the developer community a combination of products, services and benefits to produce robust and exciting applications and bring the best of them to market. The program included: Software Development Kits (SDKs) and tools, application testing and certification, technical support (both onsite and online), training, developer services, networking events and an interactive web-based community/portal to support developers worldwide.Applications that were marketed through the MAGNet programme were tested by independent test agencies against predefined criteria and evaluated as best in class. Motorola endorsed this approach as a way of ensuring high quality, safe and reliable software products in the increasingly open world of wireless application development. The company that birthed the cellular / mobile industry was a great visionary indeed !!

SK Telecom's venture capital arm launched SKTA Innovation Accelerator with a view to seed start-ups in core technologies such as data center and IT, in what the operator cloud technologies and software.16 Partnership is a key focus of the initiative, with SK Telecom Americas pairing entrepreneurs with strategic advisors as part of its value-add. Meanwhile, in July, Japan's KDDI announced the launch of its latest US$50m fund, Open Innovation II. At the same time, it announced that it had invested US$8m in four US start-ups — an educational social network (Edomo), a free digital publishing platform (ISSUU), a seat upgrading service (Pogoseat) and technology news media (VentureBeat). Fellow Japanese operator NTT DOCOMO already has a US$109m fund in place, while Softbank created a US$250m fund in 2013.

In February last year , Telecom Italia announced a €4.5m fund targeting the most innovative players in digital, mobile and green information and communications technology (ICT). The move forms part of a broader initiative, complementing its Working

Capital Accelerator that has funded 19 start-ups and assigned 109 grants in Italy over a 5 year period. One of the most interesting features of the current wave of accelerator initiatives is how operator needs are shaped by local market needs and competitive dynamics. Asian operators have been bolder than many of their peers in terms of the scope of their capital arms' most recent initiatives.

A Safaricom (Kenya) fund has been launched for an initial period of two years and will offer equity and other debt solutions of USD75,000 to USD250,000 to start-ups. Eligible start-ups must have a functioning product or service, an active user base, a team in place capable of achieving goals presented, and be based in Kenya. The establishment of the fund shortly follows Safaricom's partnership with Dynamic Data Systems to launch M-Pesa payments tracking mobile application M-Ledger.

Telekom Malaysia (TM) is to launch its first startup accelerator programme in January 2015 in a move that aims to address two problems with a single solution. The country's largest telecom and ISP giant wants to find innovative solutions to problems within its own verticals and at the same time help nurture startups in Malaysia's ecosystem. TM, with its 2.3 million fixed broadband customers and 500,000 SMEs, is looking for a way to stay relevant in its market through its new Innovation Exchange, but at the same time wants to position itself as a player in Malaysia's emerging startup ecosystem that is taking on an ever-larger role in Southeast Asia.

TM is spending around US$1.8 million to US$2.3 million on creating solutions for its verticals. But the development time is slow and often the company discovers only after the spend that it doesn't require the whole suite of services developed. This is where it hopes to bring in innovative startups. One such example is Ebizu, a startup that TM is currently working closely with. It is a point-of-sale system provider that was just a small startup when

TM discovered it. Now TM is packaging Ebizu's product together with its SME package known as 'Shop In A Box', which offers a range of tools for merchants. No doubt that it's important to be in contact with disruptive innovators creating products that Telcos might want to sell to its customers.

Bottom Line : In the era of declining voice profits Telcos should be looking for innovative products to bundle together with their core offerings .Telcos need to deepen their relationship with startups so that they can have small experiments and pilots with the products and services which would add value to their business. Actually the philosophy behind an Accelerator program is predicated on savvy thinking and the inclusivity mindset. A mindset that says I cannot develop disruptive new innovations in house because I am too busy ensuring the integrity and quality of my network : so lets outsource this activity to people who enjoy this apps / platform development game and still keep some control.

As pointed out by Ey , a holistic innovation program can leverage accelerator programs can also be taken, from opening APIs to key platforms and revisiting employee incentives, to realigning in-house capabilities with local market needs and opportunities. Nurturing and acquiring start-up capability is more than a "nice to have" for many players; it is a mission-critical route toward new capabilities in an age where fresh approaches to service creation have never been more valued and improvements in time-to-market are vital.

6.3 : Growth Perspective : Guru Insights on mHealth

The GSMA believes that mobile operators, through their evolving capabilities in creating meaningful connections between people, organisations and ultimately systems, can have a dramatic impact on the healthcare industry in improving access, reach and quality. Today the GSMA (Invitation only) Ministerial sessions on m health certainly highlighted that developed countries are under intense pressure to reduce the cost of healthcare, whereas developing countries have a greater need to deliver life-saving services more broadly.

The session deliberated on several key questions : What role can mobile play in delivering future healthcare worldwide? How can governments embrace the significant opportunity to improve quality of life for individuals, and increase access and efficiency in health services for its citizens? What are the emerging best practices from mHealth implementations. Believe it or not TB is a largely curable disease but requires six months of diligent adherence to the medication regime. mHealth could help control TB mortalities by ensuring treatment compliance through simple SMS reminders. Delivering mobile-assisted awareness to pregnant mothers and traditional birth attendants could reduce perinatal and maternal mortality by 30%.

In some developing countries, where there is less or totally non existing technical infrastructure and eHealth systems installed in hospitals, so the focus will be on basic health data collection, basic ICT infrastructure such as connectivity, and health access. Clinical adoption issues (relative to the developed world) will be lower, although the degree of IT literacy will cause issues depending on the specific market. However, budgets will also be correspondingly tighter, particularly in relation to eHealth systems which have been developed to cater more to the Western market, and particularly as health budgets will be devoted more to basic health provision, medical supplies and manpower.

As mobile operators develop capability in the ICT space and begin to replicate the capabilities of an ICT infrastructure provider (such as IBM, EDS and Oracle). This may be acceptable in markets where the mobile operator has a natural incumbent advantage – but in other markets where this is not the case the key challenge then becomes one of differentiation.There are two parameters that provide mobile operators with a unique value proposition in the eHealth industry according to GSMA :

1. Leverage Global Integration capabilities for Supply Chain efficiencies

Traditionally, the mobile operator value chain consists of those core capabilities that enable it to acquire customers through its sales and distribution network, set them up on the network, identify and connect consumers on a network, create value added services in both voice and data, provide customer service and run sophisticated billing and tariff programs to optimise revenue per customer. In recent years, some mobile operators (particularly those with strong group operations) have developed specialised global business integration capabilities, ranging from cloud computing, portal technologies, payment mechanisms, Machine-to-Machine (M2M) platforms and solutions, and systems integration. These are the capabilities that allow the mobile operator to create solutions connecting the healthcare providers with the patient, as well as with other healthcare players in the industry, providing the industry level integration.

2. Shifting Demand with Integrated Participative Healthcare

In developing countries, the chronic shortage of healthcare professionals as well the prohibitive cost of building healthcare facilities in rural areas, are also indicators of a need for healthcare to develop beyond its traditional hospital-bound model. This can be done either by keeping patients from entering the system through effective prevention and wellness, or by managing patients consistently after they exit the system through medication adherence, effective monitoring and post-discharge management, particularly in the case of chronic diseases.

One key opportunity discussed is the ability to take mobile operators' core capabilities and apply them inventively to solve healthcare problems both large and small. Telefonica talks of a "lift and shift" economies of scale methodology whereby there is a rule about being able to re-use a significant percentage of the investment made into any particular technology in another region.". AT&T speaks of "utilising all of AT&T's core assets to apply to healthcare", and also of the importance of "scaling up" applications in both simple and complex settings, working from small to large scale.

Contingent upon the mobile operator's global business integration capabilities is its ability to offer health specific solutions e.g. Cloud-based PACS or records hosting, remote monitoring solutions to manage chronic diseases, or sophisticated tele-medicine capabilities incorporating collaboration technologies with remote diagnostic equipment. These will enable the mobile operator to differentiate its offering while leveraging on its core ICT capabilities. Depending on the mobile operator's business strategy, such capabilities can be developed either through internal development, partnering or acquisition. This strategy however requires greater internal investment in order to develop specialised health expertise, as well as to select suitable business partners with which there is mutual benefit.

Collaboration is an example of a capability that can apply both within the enterprise, between different healthcare institutions, as well as between the institution and the patient. At the simplest level, this can be web-enabled audio and video-conferencing solutions between physicians, to sophisticated solutions which incorporate both videoconference as well as peripherals which allow specialists to conduct diagnostic assessments to patients in remote areas. Cisco, for example, partners with both Telefonica and AT&T to provide such solutions. In Telefonica's case, the Health Presence product was introduced in order to alleviate the travel needs for patients in the Balearic Islands who previously had to travel to Mallorca for diagnosis.

Orange set up their Orange Healthcare business unit in 2007, talks of "joining up healthcare", echoing the role of mobile operators in connecting both the healthcare enterprise and the patient, as well as helping to remove the boundaries between the increasing number of players in the healthcare industry. They have organised themselves into services supporting the patient (in terms of wellness and prevention), to services supporting the healthcare professional and hospital operations, to services connecting the two in order to better manage the number one problem driving healthcare costs globally, that of chronic disease management.

Telefónica, which set up their global eHealth unit in 2009, speak of their unit as being a "standard bearer for health products and services", with three focus areas in Remote Patient Management (RPM), Telecare and Health IT. Grounded in the belief that healthcare both has a local context, as well as a need to cross-pollinate ideas and scale across geographies, their organisational strategy was to have separate individuals which have both a functional responsibility, as well as a regional/country focus

While the mobile operator industry relies a lot on partnerships to get into the health industry and expertise, the more established players do recognise the need to create their own in-house health expertise. When AT&T started their health business unit, for example, it hired 60% of their staff directly from the healthcare industry. Telefónica decided to build their team combining staff coming from the healthcare industry with experienced employees with deep knowledge of ICT as well as of the functioning and capabilities of a telco operator. Orange has 100 R&D professionals in healthcare in 4 Skill Centres globally in order to develop expertise tailored to healthcare systems in their global footprint. Telefonica centralises their technical resources in a Living Lab in Granada which houses all their healthcare application development expertise.

Cloud-based eHealth systems are gaining traction, as the number of locations of healthcare practice increase, along with the high cost of deployment and maintenance for traditional client-server models. These are particularly attractive to smaller clients with less complex needs, or clients with budget constraints. However, there are still residual concerns with the security and reliability of using a fully cloud-based solution for mission-critical health applications. Security and reliability can often be obtained through more "private" clouds which offer dedicated resources and guaranteed access, which unfortunately reduces the core cost advantage which cloud systems are procured for.Use case for the cloud is in the hosting of health records or acting as a connector between different enterprise level EMR systems. AT&T's Healthcare Community Online product is a health information exchange based on cloud technologies, supporting collaborative care through secure messaging, access to multiple applications through one portal, integrating with the American Medical Association's own portal.

Be sure of one thing " the mobile operator's evolving participation in eHealth will ultimately depend on the extent to which there is mutual value creation between the two sectors. For the mobile operator, there is the promise of a new corporate segment, a means to better utilise and monetise their enterprise ICT capabilities, an opportunity to extend their brand in the healthcare industry, and provide more subscriber value in their own mHealth services as they better integrate into the enterprise healthcare sector. SMART, a mobile operator based in the Philippines, recently rolled out a lightweight eHealth system in two major cities, with these constraints in mind.

A mobile operator based in South Korea, has also recently developed a hospital information system that it plans to roll out across various sites in its home country, and currently considering expansion to China.For the healthcare sector, the mobile operator's involvement represents an opportunity to reap the large economies of scale of using mobile operators' significant IT investment, and partnering with an ICT player who is best placed in helping it extend its reach to the patient.

---------------------------------♠---------------------------------

6.4 : Security and IDM : the hidden gold for Telcos in the smartphone mine

Privacy and the integrity of personal and corporate data is becoming one of the most significant areas of the digital world, driven by trends such as BYOD increasing the use of cloud computing in the corporate space, or the continuous use of social media in the consumer space. As online fraud, e-crime and many emergent threats move into the tablet and smartphone space, a solid commitment to offering products and services to protect customers and their digital lives becomes paramount for any self respecting Telco.

Welcome to the dire need for Identity Management (IDM). In a nutshell, IDM is an integrated system of business processes, policies, and technology that lets an organization control user access to online applications or services while protecting the user's privacy and the organization's resources.Telcos that undertake strong IDM service programs can capitalize on their existing assets to create additional revenue streams with subscriber identity data.

In 2012, the Edelman Trust Barometer indicated that distrust amongst consumers was growing; compared to just a year ago, twice as many countries surveyed are now skeptical and fewer countries are neutral. In fact, a privacy survey of 9,200 interviewees in 14 countries around the globe conducted by the research company Psychonomics reveals that customers are increasingly concerned about the use and misuse of their data. 82% of respondents see privacy as an important topic; 76% are concerned about privacy violations; and 45% of subscribers feel they lack control over their personal data. Only 9% of respondents don't care about data protection in general, whereas 80% of respondents claim to be selective about sharing their personal data and why. A further 73% are afraid of their personal data being sold to third parties. In a digital age characterized by complex¬ity and concerns about privacy and security, customers are increasingly drawn to the idea of a Trusted Identity Provider.

However, the major challenge that operators face in using their data to generate revenue, is that customer data is still siloed on legacy systems, due to their past merger and acquisition activity. Therefore, operator IT departments would need to invest in data integration, MDM (Mobile Device Management) , data quality tools and the necessary in-house expertise to use these tools effectively. If Telcos could leverage this subscriber data in real-time they can not only improve their service offerings, but can also create completely new revenue streams and support third parties in numerous industry sectors in improving their service offerings to their customers.

Before Telcos can become true and trusted identity providers, they need more than just authentication, authorization, and accounting (AAA) servers inside their networks; they also need directory servers, tools for managing subscriber access, and the federation software itself. Thats big investments but if the Tecos don't do it the Web 2.0 players will do and capitalise on this huge IDM opportunity. Most executives compare the cash flows from innovation against the default scenario of doing nothing, assuming—incorrectly—that the present health of the company will persist indefinitely if the investment is not made. For a better assessment of an innovation's value, the comparison should be between its projected discounted cash flow and the more likely scenario of a decline in performance in the absence of innovation investment.

Some Telcos may not really know yet how they will make money as an identity provider, but they are certain that IDM will help deliver services that can reduce customer churn. Fortunately proactive operators do see this opportunity and are acting on it. France Télécom – Orange is among a handful of European and Asian network operators that are implementing IDM technology and learning to become identity providers – companies that can, in effect, vouch for a user's identity in transactions with Web-based service providers. In Argentina, operator MoviStar, is offering a single sign-on service. With one single password, the user can access to online services such as Facebook, Twitter and Flickr moving to "zero sign on", with all a customer's identities stored on the secure element of the SIM card.

Telfonica has segmented its Security market under Information Security , Electronic security and Mobile Security mapped across Consumer / SME and Enterprise / Gov sectors and sales estimated at $ 300 million. Their Enterprise security solutions include Web security gateway (secure internet surfing) , Clean e-mail (mail security service) , Anti-DDoS(preventing Distributed Denial of Service attacks) , Anti-phishing and fraud prevention and Managed Security services. The Device Management Service is available in a tiered 'pay as you use' structure incorporating:
• Device inventory: know what you've got and where it is
• Application management: update, remove and review software across devices
• Capability control: enable or disable functions or capabilities on devices
• Advanced security: keep your data safe with encryption, password management and firewall

The good news for operators is that it is not too late to join the race.A Telco's success in generating revenues as Identity Brokers depends on upon a close alignement with Enterprise verticals such as Healthcare , Finance and Goverment. With their many assets, operators are in a very good position to take a leading position in the Internet value chain. The operator can also act as " Trusted Identity Guardian " for end users and as part of a larger commercial ecosystem.

---♠---

6.5 : Mobile Money : much more than a gravy train !!

Every Mobile Money Congress attracts an eclectic group of high profile people from the Banks , Telcos , NGO's , Govt , Vendors to discuss the ramifications of Mobile Money for Africa. For the uninitiated a Mobile Payment platform is a piece of complex middleware (aka Transaction Engine) that is built on Java to facilitate transactions on mobile phones using SMS , IVR , USSD or WAP. It sits like a glue on hardware servers to manage accounts, offers optimal user interfaces, processes transactions, and provides the full suite of resources required for a mobile money services. Sometimes it is used as a payment gateway to bring mobile money to an existing financial service.

The mobile payments marketplace is a complex one. The range of players involved in the mobile payments ecosystem; its integration with the online world; and its role in both remote and point-of-sale (POS) payments all suggest that the marketplace will continue to evolve rapidly. Mobile money displays the characteristics of a platform bringing together financial services providers and clients, and providing them a core functionality which they can use to transact, and which can be incorporated into different financial products. As a network infrastructure, and as a platform for financial services innovation, mobile money appears will radically reconfigure how retail finance is done in developing countries.

GSMA estimate s that overall, mobile payment services are expected to reach US$245b in value worldwide by 2014. At the same time, mobile money users are expected to total 340m, equivalent to 5% of global mobile subscribers.To date, many of the most successful services worldwide have originated in developed Asia — where mobile contactless services are well established — alongside money transfer services in emerging markets such as the Philippines and Kenya. Smartphone penetration allied with a more mature e-commerce market is producing a new wave of innovation in handset payments. Mbanking is predicted to reach 500million users in the next three years.

Bearing in mind that this number currently excludes the 2.5billion unbanked consumers with no access to traditional financial infrastructure, the actual figure for penetration levels by this time is likely to be far higher. Indeed, today's 141million mpayment users are only the start with mobile transactions predicted to grow to $1trillion globally by the same year. Even the green shoots of mCommerce can be seen, with 500million people already using mobile devices as transport tickets and over 863million NFC enabled phones expected in circulation by 2015.

Banks and financial institutions launching mobile wallets in the immediate future should expect to enter a hotly contested market, crowded with own-brand solutions that are limited to the delivery of proprietary services only. The majority of financial institutions will, in the short term, attempt to develop their own proprietary wallets in a partner-independent manner as we have seen in South Africa. Only when a secure element (SE) is required, or when core functionalities become too difficult for financial institutions to achieve alone, are they likely to open their solutions and seek to cooperate with other stakeholders. The number, breadth and variation of mobile wallet solutions set to come to market is going to make getting to grips with the technology a challenge for the end-users. This means that banks and financial institutions should think very carefully about their chosen structure and approach to market.

Most people in the know feel strongly that mobile money will become a "catalytic platform." The variety of new models and approaches being tried could portend a fairly fundamental realignment of the cash-based financial sector moving from all cash transactions mediated by expensive retail infrastructure to greater use of electronic payments through cell phones. Outsourcing cash handling will not only allow financial service providers to serve their clients at lower cost per transaction but also allow them to get more value out of their existing front office infrastructure and staff as they focus on more sophisticated tasks such as customer service, cross-selling, risk evaluation, etc. as opposed to cash handling.

On the client side, customers will gain access to a dense network of transaction points, greatly reducing their costs to access financial services. And once clients are on the financial system, and able to transact at low cost with financial service providers, the platform enables a whole new set of services and delivery models which were not previously possible or profitable.

Mpayments will continue to flourish as trust and confidence grows – firstly for traditional transactions, such as bill payment, and then eventually facilitating mobile unique applications such as transfers using mobile phone numbers or NFC proximity payments. As Mobile Money matures, mCommerce services will rapidly transform the consumer experience through seamless ecosystems delivering new levels of service, convenience and satisfaction. The consumer revolution and increasing competition will fuel the growth of Mobile Money from complementary channel to strategic keystone, delivering connected, contextual services that were once unimaginable but now highly in demand.

Governments and regulators must decide on the appropriate policy framework for mobile money services. Much will pivot on whether mobile money is considered an extension of existing payment mechanisms or a unique channel. Proportionate regulation and cross-industry incentives have an important role to play. Operator-led money transfer services have gained traction due to factors such as the high penetration of handsets compared to bank accounts and benign regulatory environments. However, opportunities have yet to be exploited in many developed existing payments channels. In these countries, the mobile channel is more closely connected with the development of electronic payments, while the disparity in coverage levels between banking and mobile infrastructure in emerging countries gives the mobile channel a more singularly transformational role.

Mobile Money is high volume low margin game. We are talking 2 to 3 % per transaction not 10 to 20 %. Since mobile money is about financial inclusion a high per transaction fee excludes the unbanked at the bottom of the pyramid. Understanding and segmenting the target market is critical before rolling out a service. Payment solutions must be carrier grade , scalebale and secure with the functionalities desired by the end user. A motivated Agent network is vital portion of the value chain.

6.6 : Monetizing IP MPLS services : Strategic Parameters !!

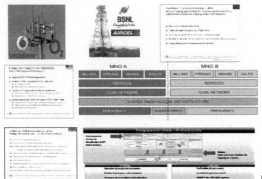

MPLS VPN is a family of methods for harnessing the power of Multiprotocol Label Switching (MPLS) to create Virtual Private Networks (VPNs). MPLS is well suited to the task as it provides traffic isolation and differentiation without substantial overhead. Healthy growth in the market is expected as organizations increasingly look to IP MPLS VPN services to control WAN costs in the face of rising bandwidth needs, to Ethernet for the lowest cost-per-bit, and to managed layer 2 and layer 3 services, where the expertise, knowledge, and tools of service providers can help stabilize WAN costs and prioritize critical applications while increasing capacities

Global Ethernet and MPLS IP VPN service revenue grew a combined 13% in 2011 to just over $50 billion, fueled by surging data traffic, cloud services, and cost-cutting initiatives. Asia Pacific is expected to overtake EMEA as the leading region for IP MPLS VPN services in 2013 and already leads in Ethernet services; Asia will remain the leader for the combined IP MPLS VPN and Ethernet services market going forward, led by China and India. IPsec VPN services accounted for the remaining share of 32.5%. In Middle East and Africa, MPLS VPN services accounted for largest share (67.5%) within IP VPN market in 2012.

Concurrently carrier Ethernet Exchanges are a significant new development that facilitate Ethernet connections and accelerate the move to Ethernet transport and services. In these exchanges service providers pay small fees to a Carrier Ethernet Exchange to make it easy for them to locate, buy, and provision Ethernet connections from each other. This in turn jumpstarts more Ethernet services and more of the IP VPN services that ride on Ethernet transport. The net effect of these new Ethernet exchanges, combined with fast-rising mobile backhaul connections, is a quickening of the Ethernet and IP VPN services markets

To capitalise on this growth opportunity BT Global Services invested in the rollout of new IP MPLS infrastructure, services, and expertise into Turkey, the Middle East, and Africa. On the face of it, the Middle East and Africa is a sprawling super-region, with as diverse a range of markets as could be found on the rest of the planet. The startups and traders of the narrow streets of Nairobi are a world away from the huge family-owned conglomerates of Turkey or the state-owned petrochemical giants of Saudi Arabia. Yet these enterprises have important things in common: they are part of the same global supply chains in manufacturing, transport, and logistics, and they need high-security broadband access and hosted applications. That makes the MEA a perfect target for network operators with the capability to combine their own UC platforms with those of vendors, and to provide both remote access and IT support

MPLS VPNs offer the ability to prioritize applications such as VoIP by class of service (CoS), create and improve disaster recovery infrastructures, utilize a fully meshed infrastructure that replaces outdated hub and spoke architecture, and reduce complexity to simplify network management in an increasingly complex landscape. The desire to move toward a more cost-effective network that supports voice, video and data is among the primary drivers behind the move to MPLS VPN services, with a number of other technological and financial drivers drawing enterprises in that direction as well.

All businesses need communications networks which accurately mirror their own changing data flows and their own transactions with customers. Enterprise clients need to connect site to site and function to function: the network infrastructure must enable multiple departments including sales, marketing, production, logistics and distribution to work together as a single business machine. And they want to do so in a way which minimizes cost and maximizes reliability.Among these are:

- Class of Service (CoS): provides ability to prioritize applications, such as VoIP
- Automatic redundancy/disaster recovery: create and improve disaster recovery infrastructures
- Fully meshed infrastructure: replace outdated hub-and-spoke architectures
- Reduced complexity: one network platform supports all application traffic—including VoIP and data applications

IP VPNs are typically utilized by organizations operating out of multiple office locations that require a secure, flexible and cost effective means for their employees to communicate and share information across a central computer network. MPLS networks form a proper foundation for a number of business critical applications including: VoIP phone service + Centralized merchant transactions + Remote application access (Citrix) + Remote user access + File transfer/sharing + Video delivery + Secure access to internal software applications + Outsourced network management + Compliance requirements + Central data storage & backup etc. Any business operating across multiple locations can benefit from a MPLS wide area network.

With the emergence of VoIP phone service, private wide area networks are serving a very important role in ensuring call quality, performance and security across the enterprise in several verticals such as :• Retail Businesses • Restaurant Chains • Hospitals and Clinics • Doctor and Dentist Offices • Call Centers w/ Remote Staff • International Corporations • Financial Institutions • Government Entities • Hotels • Franchises

Internet Protocol/Multi-Protocol Label Switching (IP/MPLS) has grown to become a foundation for many mobile, fixed, and converged networks. In mobile networks, IP/MPLS consolidates disparate transport networks for different radio technologies, reduces operating expenditures (OPEX), and converges networks on a resilient and reliable infrastructure, while supporting evolution to Long Term Evolution (LTE) and Fourth-Generation Mobile Network (4G) technologies.

BT seems to be taking the right approach by promising Enterprises with an extended infrastructure based on subsea cables in the region, new fiber connections into South Africa, and greater domestic connectivity. New MPLS nodes in Oman will extend the global reach of MPLS, while new NNIs (network-to-network interface) will take services out of South Africa into 12 other countries. In addition Ethernet managed services will be offered in four countries and a center of excellence for satellite established in Turkey. MPLS-based IP VPN and IPsec VPNs will be standard, accelerating Ethernet into 20 cities. This will be supported by portfolio expansion, with 10 launches in each center combining regional and global products with some local integration. A key example of this is global inbound services. BT managed security services will be available in all countries.Being able to deliver on-net access in close proximity to a customer's local offices, subsidiaries and partners all contribute to an eff ective international IP/MPLS strategy and can result in a carrier being more competitive with regard to faster local provisioning and troubleshooting, and more cost-eff ective circuit fees.

According to Current Analysis GNT the ability of an operator to provide a geographically comprehensive on-net IP/MPLS footprint that can support a customer's key sites is critical for customers when procuring global network services.Unlocking the various revenue streams out of a carrier grade IP MPLS network requires a highly tuned understanding of enterprise needs and PoP location. Being able to deliver on-net access in close proximity to a customer's local offices, subsidiaries and partners all contribute to an effective international IP/MPLS strategy and can result in a carrier being more competitive with regard to faster local provisioning and troubleshooting, and more cost-effective circuit fees.

Carriers with strong in-region MPLS PoP ownership should certainly present to customers the advantages of working with a provider that can more effectively monitor and manage its own network for better performance and rapid response and troubleshooting. Ongoing efforts to expand networks and invest in PoPs in new regions, Ethernet VPLS expansions, new Ethernet and DSL capabilities and new NNI agreements will no doubt bolster the image and credibility of the MPLS provider.Naturally, having the highest MPLS PoP count is not enough and the service wrap, pricing and SLAs are equally important; carriers must recall that Vanco, as a VNO, managed to win customers fairly effectively until it was bought by Reliance Globalcom in 2008.

The main point here is that operators must be competitive on price and provisioning times and reassure customers that they are skilled in managing and selecting local network partners and that the partners they have chosen are reliable, financially stable and can deliver the target SLAs and required resiliency for a customer's sites.

6.7 : Wearables : Worthless fad or game changer ?

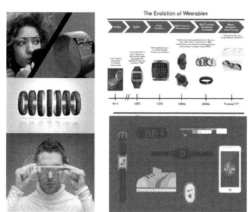

As differentiation in the smartphone space becomes increasingly difficult, top-tier vendors are looking to wearables as a way to bolster their positions – and generate additional revenue on top of often low-margin handset sales . With about 35 m wearable devices in 2013 wearable technologies are about to cross the chasm and generate sustainable growth in the years to come. The ecosystem of widely adopted smart phones, enabling technologies such as low power sensors, Bluetooth 4.0, highly integrated M2M modules, and cloud computing, as well as new app stores for wearables is getting into place. Innovative companies and start-ups across the globe are nurturing this new megatrend. Big brands like Apple, Samsung, Google, Nike and carriers will further push mass market adoption. Major players from other industries like healthcare, automotive, entertainment etc. are about to enter as well.

Wearables are quite a new phenomenon in the mainstream and in contrast to making people curious about new technical solutions it is much harder getting them to adopt wearable technologies on an ongoing base. Maintenance is another factor that decides over long term customer satisfaction and in the case of wearables, having to charge the batteries is one of the hurdles that make or break a relationship. Wearable technology is related to both the field of ubiquitous computing and the history and development of wearable computers. With ubiquitous computing, wearable technology share the vision of interweaving technology into the everyday life, of making technology pervasive and interaction friction less.

Despite the buzz surrounding Google Glass last year, generally the market appears to have settled on two preferred form factors: smart watches and smart bands (with some products moving toward straddling both). The former is not a new category, although it has hardly been overburdened with success so far. Perhaps the most interesting development in this space is Samsung's choice of Tizen to power its Gear devices – enabling it to claim continued support for the platform, while not being so hamstrung by its lack of apps. Currently, sports and activity trackers account for the largest chunk of wearable technologies shipped today. Smart activity trackers are widely available, and the device's trendy and stylish appearance makes them very popular with a broad range of customers. It is estimated 61% of the wearable technologies market is attributed to sport/activity trackers in 2013.

The increased presence of vendors such as Sony, Samsung, LG and Huawei in the wearables space could create challenges for specialised players such as Pebble (smart watches) or Fitbit (smart bands) which, while having first-mover advantages, are now faced with competition from companies which can potentially deliver scale quickly. There are also challenges in driving long-term engagement with wearables, particularly with regard to smart bands. While these devices are proving appealing to the fitness-conscious, there is little evidence that these can support additional use cases which could drive wider adoption.

Sony's Lifelog app is a good first attempt, and vendors really need to look to drive innovation in supporting apps in order to appeal to a wider customer base.Smart bands benefit from the fact that they are generally being positioned as smartphone companions, meaning much of the intelligence is offloaded onto the senior device – enabling a lower bill of materials, and therefore lower price points. But there are a number of challenges to overcome. Battery life in some first-generation products is poor, and in many cases the hardware design is not appealing enough to become a mass-market proposition.

And it is also unclear if there is scope for operators to benefit from the growing interest in wearables. Operators have invested a lot in machine-to-machine communications that go beyond connectivity to look at service enablement platforms in order to add new services, such as healthcare and connected cars.Wearable technologies will have an important role in all these new service areas. In order to exploit these opportunities, commercial and innovation partnerships are fundamental to bring new ideas to the market and test potential business models.

Telefonica's partnership with LG, Sony and Samsung will be interesting to watch, as it looks to ensure the operator is not sidelined in the new product category.The spectrum of partnerships includes the entire wearable technology value chain. For example, collaboration with specialised product solutions providers, such as medical device manufacturers or fitness device designers, can help operators with their offers. But, collaboration with fashion and design brands may become more relevant in order to get closer to consumers.

Apple has a great history in design and for its new watch the company paid close attention to build a device with a look that fits their customer's personal style Similar to its competition, Apple decided to take advantage of a colorful and high-resolution display that will deliver a great experience and also lead to a frequent use of the watch throughout the day. Although the frequent use mean the watch's battery life will be limited, charging it is easy with the magnetic, inductive connector. Even if this design decision sounds simple, it's the small differences in usability that have been important for Apple's success over the last decades.Industry pundits sees the Apple Watch as "a product that can dramatically increase the adoption of smartwatches, because of its completely new approach for user interaction increasing user satisfaction.

Since wearable computing devices let users go hands-free, there are a lot of ways they could be useful at work. For emergency personnel, search and rescue teams and mobile warehouse workers, wearables can provide high-tech mobility and tracking features. Smartglasses could be useful for technicians who need to consult a manual or a set of schematics while performing repairs. Wearables may also be able to remotely manage equipment, such as machinery on an assembly line, making the workplace safer for employees.

Workers who need to wear special suits, such as environmental disaster teams, could have hands-free access to data via smartglasses or a smartwatch. Any user who needs instant access to important data – members of sales teams, real estate agents, lawyers, rural doctors, law enforcement and fire fighters, military personnel and more – can benefit from using wearables in the workplace.

Wearable computing devices haven't gone mainstream yet, so it's hard to say whether they definitely will or won't end up in the enterprise. If you consider the rise of smartphones and tablets a harbinger of wearables' trajectory, then they could be in your office fairly soon. But consumers are fickle and difficult to please. If wearables don't take off with consumers, they may never make it to the enterprise. If users do start wearing smartglasses and smartwatches to the office, it could pose some problems for IT.

As it stands, you wouldn't ask a user to relinquish his Rolex so you could configure and secure it, but you might face that exact situation with smartwatches. Undoubtedly, there will be pushback from some users. Plus, so little is known about how wearables will work that it's tough to say if or how you'll be able to manage them. Will you be able to apply configuration profiles or mobile device management settings? How will you keep corporate data from leaking from someone's smartglasses? There aren't answers to these questions yet.

The M2M technology that enables communications between wearable devices and the end points connected to the Internet will play a key role in boosting the uptake of wearbales. The truth is that beneath the plastic covers –whether sober or colourful–of these accessories lie great and sophisticated data solutions.The heart rate, the calories burnt, the number of steps taken each day, the pulse, how fast certain movements are carried out... all that local information is built up and can be uploaded to the cloud to go through an analysis system. Data can be analysed individually and compared with the user's own history, or else they may be assessed collectively. Metrics help in setting personal goals, being more physically fit, eating more healthily or just knowing whether our health is within average.

There is also a strong "verticalization" and specialization on the market, different hardware pieces focused on sports, health, content.The key is to provide very well defined value proposition instead of a generic monitoring. The challenge is integration. Being able to mesh together all the huge amount of information received from these devices in a central interface. Here we talk about the semantics of data. I want my smartphone, my sports band, my internet hub and other relevant information like how much I sleep or what I eat all mesh into a semantic that allows to help me gain insight into leading a healthy life.

In any case wearbles will be big business poor battery life or not. By 2017, Berg Insights forecasts that 64 million of these gadgets could be sold, vis-à-vis 8.3 million sold in 2012. In turn, Juniper Research raises the first figure up to 70 million gadgets, while the ABI Research consulting firm foresees that by said date there will be almost 170 million wireless sensors available in the sports and health sectors.

Manufacturers are emphasizing health and fitness wearables because the one area where there's been some headway in organizing the tremendous amounts of information from all the sensors in the wearables is fitness and tracking.According to Research Analysts "It's a good foundation for things to come provided Telcos can bundle in wearables into their M2M solutions roadmap"

---------------------------------THE END---------------------------------------

Printed in Great Britain
by Amazon.co.uk, Ltd.,
Marston Gate.